gta papers

ARCHITECTURE/MACHINE
Programs, Processes, and Performances

Moritz Gleich and Laurent Stalder (eds.)

gta Verlag

Introduction
Laurent Stalder and Moritz Gleich

Laurent Stalder is Professor at the Institute for the History and Theory of Architecture (gta) at ETH Zurich.

Moritz Gleich is Research Assistant at the Institute for the History and Theory of Architecture (gta) at ETH Zurich.

If a machine is an "apparatus using or applying mechanical power and having several parts, each with a definite function and together performing a particular task" [1] then it stands to reason that a building is not a machine. Nevertheless, it need not be pointed out that the machine has long since carved out a successful career for itself in architecture. A number of architectural historians have referred over the last 40 years to that brief passage in the *Encyclopédie d'architecture* in which, probably for the first time ever in an official context, which is to say, in a specialist journal, mention was made of architecture as a machine. [2] Adolphe Lance, an architect and then chief editor of the *Encyclopédie*, used the metaphor in his lengthy critique of Léonce Reynaud's *Traité d'architecture* in order to draw attention to certain machinic characteristics of architecture, which, he claimed, had hitherto been neglected in architectural theory, despite a growing awareness among the general public of how science and industry were profoundly influencing architectural practice. [3] That this omission was soon to be remedied is common knowledge. In the work of architects such as Eugène Emmanuel Viollet-le-Duc, Le Corbusier, Richard Buckminster Fuller, Alison and Peter Smithson, Kishō Kurokawa, and Rem Koolhaas, the term not only outlived historicism, modernism, and postmodernism, but also swept across North America, Europe, and Japan, and thereby anchored itself firmly in the vernacular of the last 150 years. The writer Émile Zola used it with the same ease as later the cultural historian Lewis Mumford, the psychoanalyst Félix Guattari, and, more recently, the philosopher Peter Sloterdijk. In the meantime, it appears that we have come to be surrounded by rather too many than too few machines—machines for living, concrete and abstract machines, zero-degree machines, emotion machines, defective machines, desiring machines, infernal machines, or energy machines.

If the topic of architecture/machine is addressed here, then not only in order to gain clarity within this diversity but also, and not least, with a view to forging new prospects in a dual sense: on the one hand, to broaden the cultural-historical and theoretical parameters of contemporary thought on technology in architecture such as is conveyed by the term machine, and thus also to radically challenge its increasingly alleged technocratic determinism; and, on the other, to trace the history and theory of architecture back to one of the major issues in architectural practice, the far-reaching implications of which have with few exceptions been largely ignored in recent years. This is all the more astonishing given that anyone seeking to sum up the architecture

1 *The New Oxford Dictionary of English* (Oxford: Clarendon Press, 1998), p. 1108.

2 Cf. François Béguin, "Savoirs de la ville et de la maison au debut du 19ème siècle," in Michel Foucault (ed.), *Politiques de l'habitat (1800–1850)* (Paris: CORDA, 1977), pp. 211–324; here p. 306; Manfredo Tafuri, "'Machine et mémoire': The City in the Work of Le Corbusier," in H. Allen Brooks (ed.), *Le Corbusier* (Princeton: Princeton University Press, 1987), pp. 203–18; here pp. 204–5; Georges Teyssot, *Die Krankheit des Domizils*. Bauwelt Fundamente, 87 (Berlin: De Gruyter, 1989), p. 52. In 1977, Georges Teyssot and Paolo Morachiello organized a conference in Venice that aimed under the title *Le macchine imperfette* to investigate the advent of modern architecture and its planning techniques. See Paolo Morachiello and Georges Teyssot (eds.), *Le macchine imperfette: Architettura, programma, istituzioni, nel XIX secolo* (Rome: Officina Edizioni, 1980).

3 Adolphe Lance, "Traité d'architecture par M. Léonce Reynaud," *Encyclopédie d'architecture*, vols. V/VI/VII (1853), pp.33–7/47–53/62–9; here vol. VII, p. 68.

of the last century in a single word would probably have to resort to using "machine." Several strands of this machinic history of architecture can be distinguished.

1 It is not that machines failed to feature in the architectural context prior to Adolphe Lance's statement of 1853. On the contrary, talk of machines is ubiquitous in the classical tracts on architecture. In the first volume of his *Ten Books on Architecture*, Vitruvius distinguishes three subsections of architecture—the execution of buildings, horology, and the construction of machines—before devoting the tenth volume exclusively to the last of these. There, he defines machines as "a combination of timbers fastened together, chiefly efficacious in moving great weights." Machines, Vitruvius continues, are distinct from tools on account of the number of laborers involved in their operation. Elsewhere, he ascribes mechanical status to revolving theater stages and military equipment. [4] Indeed, since antiquity, the Latin *machina* has connoted not only operational machines in today's sense of the term but also static constructions, such as scaffolding, easels, or siege towers. In fact, the latter meaning prevailed in medieval times and resonated still in the eighteenth century, when it came to include all things imposing, bulky, or purposefully laid out, such as church spires, obelisks, paintings, or even theater plays. It was only in conjunction with a new reading of Vitruvius's texts by savants and architects such as Leon Battista Alberti and Georgius Agricola that the modern notion of machines as gearing mechanisms and work engines began to gain currency around 1500. [5] In regard to construction one can therefore find Vitruvius's understanding of the machine practically unchanged in the late eighteenth century: even as the Industrial Revolution began fuelling advances in technology, manufacturing, and science, Jean-Baptiste le Rond d'Alembert spoke of machines in this sense in the ninth volume of his *Encyclopédie* of 1765: as wooden structures designed for the movement of loads. [6]

2 That the ensuing and comprehensive mechanization of the environment from the late eighteenth century on—in agriculture, transport, and manufacturing—wrought radical and far-reaching changes in architecture can be called to mind here only briefly. Architectural theory was in any case slow to respond. When Quatremère de Quincy noted in the architecture volume of his *Encyclopédie méthodique* of 1820 that "architecture, because of its numerous relations, is the art that implements the biggest number of machines," he was still thinking mainly of construction machines, [7] despite the ever-growing number of machines being

4 Vitruvius, *The Ten Books on Architecture*. Trans. Morris Hicky Morgan (Cambridge, MA: Harvard University Press, 1914), pp. 16, 283, 150, 303–19.

5 Cf. Marcus Popplow, "Die Verwendung von lat. *machina* im Mittelalter und in der Frühen Neuzeit—vom Baugerüst zu Zoncas mechanischem Bratenwender," *Technikgeschichte*, 60 (1993), pp. 7–26; Peter Friess, *Kunst und Maschine: 500 Jahre Maschinenlinien in Bild und Skulptur* (Munich: Deutscher Kunstverlag, 1993), pp. 16–24.

6 *Encyclopédie ou Dictionnaire raisonné des sciences, des arts et des métiers*, vol. 9 (Paris: Briasson, 1765), p. 798. A large part of the entry is adopted from Ephraim Chamber's *Cyclopaedia: or, An Universal Dictionary of Arts and Sciences*, vol. 2 (London: James and John Knapton, 1728), pp. 478–9.

7 *Encyclopédie méthodique. Architecture*, vol. 2 (Paris: Agasse, 1820), pp. 645–6.

put into operation inside buildings at that time: devices to supply gas or water, apparatus such as lifts or bell pulls, and means of heating, cooling, or ventilating air. It was only in the second half of the century that architectural literature began to consistently address the machine. For instance, an early volume of the *Handbuch der Architektur* of 1881, one of the last comprehensive studies of architecture to be published, is devoted to *Services to Supply Buildings with Light and Air, Warmth and Water*, that is, to the numerous mechanical systems to have permeated the building industry in the foregoing century, and thus definitively incorporates them in the architectural realm. [8]

The scope of these technical interventions was initially quite limited. Each individual device originally has a life of its own and the technical object in its early domestic manifestation may be described as abstract. It leads an independent existence and functions according to its own internal logic, regardless of its context or neighbors. Above all, it is independent of the space-defining shell construction into which it settles or whose tectonic borders it penetrates. Yet, little by little, it transforms the built structure into a mechanical ensemble, the distinct parts of which have hence to be coordinated not so much by way of compromise as on account of their simultaneity. [9] From here, it is only a small conceptual step to seeing architecture as an integral part of a universal "mécanologie," as the French architect and engineer Jacques Lafitte shows in his essay of 1932, where he defines architecture—in contrast to the "natural" or "raw bodies"—as an "organized body" and matter-of-factly assigns it to the category of (passive) machine. [10]

3 The examples and trajectories introduced above are ambiguous insofar as the borders between the machine—as a concrete apparatus, a device, or a gadget—and the machine analogy tend to blur. When Lance writes of architecture machines in 1853 he mentions neither construction equipment, nor the equipment used to survey or portray architecture, nor even explicitly the countless infrastructural innovations increasingly shaping the architecture of his day. Remarkably, he demands instead that the house itself be regarded henceforth as an instrument, as a machine, so to speak: one that not only serves to shelter man but also, whenever possible, subordinates itself to his needs, supports his activities, and multiplies the fruits of his labor. [11] As Lance suggests (and his successors' positions echo), the notion of the machine is not so much a reference to the nascent mechanization of the house itself around 1850—and the attendant systematic approach to issues of heating, cooling, and

8 *Dictionnaire de l'Académie française*, vol. 2 (Paris: Firmin Didot Frères, 1835); *Anlagen zur Versorgung der Gebäude mit Licht und Luft, Wärme und Wasser. Handbuch der Architektur*, part 3, vol. 4 (Darmstadt: Diehl, 1881). For general accounts of the technization of architecture see Pedro Guedes (ed.), *Encyclopedia of Architectural Technology* (New York: McGraw-Hill, 1979); Cecil D. Elliott, *Technics and Architecture: The Development of Materials and Systems for Buildings* (Cambridge: MIT Press, 1992).

9 Gilbert Simondon, *Du mode d'existence des objets techniques* (Paris: Aubier, 1989 [1958]), p. 21. Simondon does not describe the transformation of architecture in this passage, but the evolution of the motor. The analyzed process of individuation nevertheless seems to be appropriate to the realm of architecture.

10 Jacques Lafitte, *Réflexions sur la science des machines* (Paris: Vrin, 1972 [1932]), pp. 24–5, 67–70. Simondon resorts to Lafitte's classification in his lectures at the Sorbonne and the Ecole Normale Supérieure (1968–1969), amongst others. Cf. for example Gilbert Simondon, *L'invention dans les techniques: Cours et conférences*. Ed. Jean-Yves Chateau (Paris: Seuil, 2005), pp. 170–93.

11 Lance, "Traité d'architecture" (see note 3), vol. VII, p. 68.

ventilation—but, rather, a means to describe architecture in terms of its overall performance, which can henceforth be assessed in terms of its efficiency and planned accordingly.

The machine metaphor has a history of its own. From antiquity on, machine analogies permeate such diverse fields as cosmology, physiology, and politics, in the service of a general description of operative divine or human creations. In the Middle Ages, in keeping with the term's then primarily (yet often overlooked) static connotations, its metaphorical use most often emphasizes the stability of a composite object, be this the world (as in *machina mundi*), the body (as in *machina corporis*), or an architectonic structure. [12] It is the technical and cultural developments of the Early Modern period that pave the way to the modern notion of the machine, preparing it to serve as an image for complex, dynamic, and productive entities as well as predetermined, regulated processes—as, for example, in the now kinetic "world machine." [13] However, apart from certain singular and specific exceptions, it is only in the nineteenth century that machine metaphors begin to spread in the discourse of architecture. [14] The publication of Lance's article around 1850 may therefore be regarded as a crucial shift. The industrialized economy has so altered living conditions and machine concepts as to provide now a role model, also for architecture. [15] Yet even Lance remains hesitant. "If you'll please excuse this bizarre metaphor," he writes, "our house is the factory in which we produce the myriad acts of our private life." [16]

4 If in architecture the machine analogy emerges fully only in the mid-nineteenth century, one must inquire not simply into the exact nature of the problems it conveys, but also into the terms hitherto used by architects and historians of architecture to describe similar issues. Here too, Lance provides insight: "Might it not be possible," he asks, "to regard buildings in the light of something other than their disposition and distribution?" [17] He thus shifts the focus from the arrangement of spaces within a building—*disposition*—and the internal

12 For a general historical account of the machine notion and metaphor, see *Historisches Wörterbuch der Philosophie*, vol. 5 (Basle: Schwabe, 1980), pp. 790–802. To document the use of "stable" machine expressions in construction, Popplow cites the twelfth-century author Honorius of Autun: "It is better to have the whole machine of the house carried by one marble pillar than to spread its entire load over one hundred supporting canes." (Trans. by the authors). Popplow, "Die Verwendung von lat. *machina*" (see note 5), p. 14.

13 Cf. Popplow, ibid., pp. 20–1; Hans Holländer, "Maschinen- und Labyrinthmetaphern als Topoi neuzeitlicher Weltbeschreibung," in Holländer. (ed.), *Erkenntnis, Erfindung, Konstruktion: Studien zur Bildgeschichte von Naturwissenschaften und Technik vom 16. bis zum 19. Jahrhundert* (Berlin: Gebr. Mann, 2000), pp. 577–86.

14 In 1674 for example, the poorhouse of Genoa is described as "macchina è un Chaos" (a chaos-causing machine), cf. Eliana Perotti, "Vom 'work-house' zur 'company town': Der britische Diskurs über Städtebau, Sozialpolitik und Arbeit im Zeitalter der Globalisierung," in Vittorio Magnano Lampugnani, Katia Frey, and Eliana Perotti (eds.), *Stadt & Text: Zur Ideengeschichte des Städtebaus im Spiegel theoretischer Schriften seit dem 18. Jahrhundert* (Berlin: Gebr. Mann, 2011), pp. 47–65; here p. 49. Other instances are to be found in the context of the rational layout of fortification, where the mathematician Guiseppe Moletti writes in the mid-sixteenth century: "una fortezza si può con ragione dire machina" (a fortress may be justly named a machine), cf. Tobias Büchi, "Naturphilosophie, Mathematik und Handwerk: Buonaiuto Korini und die Analogie von Maschinenbau und Festungsbaukunst," in Bettina Marten, Ullrich Reinisch, and Michael Korey (eds.), *Festungsbau: Geometrie—Technologie—Sublimierung* (Berlin: Lukas Verlag, 2012), pp. 119–33.

15 *Historisches Wörterbuch der Philosophie* (see note 12), pp. 797–800.

16 Lance, "Traité d'architecture" (see note 3), vol. VII, p. 68 (Trans. by the authors).

17 Ibid.

division of a building in accordance with its various purposes—*distribution*—to the "économies de temps et de forces." [18] In other words, Lance turns the spotlight on architecture's capacity to engage in temporal and physical processes. Unlike in earlier writings, economy now concerns not only the building itself but also all the movement within it, the flow of which is regulated, ordered, and modulated according to the imperative of efficiency. [19] That which is first described in Lance's work by the machine analogy is thus a dimension of architecture that transcends its materiality: its capacity to perform. [20]

Lance lays down a precise technical terminology for the idea of efficiency in architecture, the origins of which can be traced back to the seventeenth century, when it crops up first in military architecture and then, later and more universally, in connection with the construction, statics, and material properties of buildings in general. [21] Yet it is in the context of late-eighteenth-century debates in France about the hospital building as a "curing machine" that its concrete architectural as well as conceptual implications truly leave their mark, for only here does the idea of the machinic extend "beyond the material fabric into problems of space allocation and the arrangement of activities." [22] In this respect the architectural realm is no exception to the broader cultural trends of the day, wherein a shift from the "substantial" to the "functional," which is to say, effective notion of the machine can be noted from around 1800. [23]

Following the publication of a landmark study in 1976, "curing machine" becomes a kind of catchall term in the architectural historiography of the eighteenth and nineteenth centuries, [24] with the tenor to suggest the juxtaposition of a traditional reading of architecture and the cold rationality of a mechanism. [25] Yet the relevant sources actually do not refer to a "machine à guérir," nor even to a straightforward "machine," but to "machines de physique" [26] —in the eighteenth century a common term for laboratory apparatus or equipment. The physician Jacques René Tenon uses this term to describe the architecture of a hospital

18 Ibid. For the terms *disposition* and *distribution* cf. *Dictionnaire de l'Académie française*, vol. 1 (see note 8), pp. 561, 566.

19 Cf. on this, particularly Jacques Guillerme, "À propos du concept de rendement," in Jean Rostand (ed.), *Actes du XIIe congrès international d'histoire des sciences*, vol. 4 (Paris: Blanchard, 1971), pp. 82–7.

20 Cf. on this Adrian Forty, who makes similar links with the term "circulation." Adrian Forty, "'Spatial Mechanics'—Scientific Metaphors," in id., *Words and Buildings: A Vocabulary of Modern Architecture* (London: Thames & Hudson, 2000), pp. 86–101; here p. 90.

21 Liane Lefaivre and Alexander Tzonis, "The Machine in Architectural Thinking," *Daidalos*, 18 (1985), pp. 16–26, and also Lefaivre and Tzonis, "The Mechanization of Architecture and the Birth of Functionalism," *VIA*, 7 (1984), pp. 121–43.

22 Lefaivre and Tzonis, "The Machine in Architectural Thinking" (see note 21), p. 25.

23 See Thomas Brandstetter, *Kräfte messen: Die Maschine von Marly und die Kultur der Technik 1680–1840* (Berlin: Kadmos, 2008).

24 Michel Foucault et al. (eds.), *Les machines à guérir (aux origines de l'hôpital moderne)* (Paris: Institut de l'environnement, 1976). A revised edition was published in Belgium shortly afterwards (Brussels: P. Mardaga, 1979).

25 Anthony Vidler, "Confinement and Cure: Reforming the Hospital, 1770–1789," in Vidler (ed.), *The Writing of the Walls: Architectural Theory in the Late Enlightenment* (Princeton: Princeton Architectural Press, 1987), pp. 51–72; here p. 61; Robin Middleton, "Sickness, Madness and Crime as the Grounds of Form," *AA Files*, 24 (1992), pp. 16–30; 25 (1993), pp. 14–29; here 24, p. 20.

26 Jacques René Tenon, "Lettre à Messieurs de l'académie des sciences de Pétersbourg," September 11, 1788, Bibliothèque nationale de France, Paris, Nouvelles Acquisitions Françaises 11357, fol. 129–30.

not in a directly utilitarian or productive context, but as part of a trial procedure, which—although serving a specific purpose—has no predetermined outcome. What he thereby expresses is not so much the idea of deterministic automation as that of an operative entanglement, in the sense of a new form of alignment between the architectural object and the epistemic procedure of healing.

5 While, in the debate on hospitals, the term "machine" applies to the ground plan and spatial dimensions and, in Lance's text, further encompasses the ability of architectural elements or installations to adapt to their user's daily actions, for Viollet-le-Duc, a few years later, it connotes the mechanical interplay of construction elements that constitute, once combined, a functional entity. [27] Le Corbusier, for his part, in the opening decades of the twentieth century, famously foregrounds the instrumental character of built objects: the house as a "machine à habiter," the chair as a machine for sitting, or the temple as an emotion machine. [28] The "air machinist" Le Corbusier conjures is characterized not so much by the particular technological stage of development of the objects of reference—for they are primarily pre-industrial artifacts—but, more emphatically, by the ubiquity of these objects in his imaginary. [29] For Reyner Banham, who first critiques this position of Le Corbusier in the 1950s, the machine eventually comes to concern the issue of climate [30] while Nicholas Negroponte shortly afterwards begins to open up a new line of interpretation in his computer-based research on "architecture machines" at MIT. [31] Kazuo Shinohara, by contrast, in his machine metaphors of the 1970s, 80s, and 90s, references Gilles Deleuze's and Félix Guattari's desiring machine, whereas Guattari develops a short theory of the architectural machine in reference to the buildings of Shinohara's compatriot Shin Takamatsu. [32]

Indeed, it is this considerable flexibility that distinguishes the machine analogy in architecture over the last 250 years or so: the comparative instances change and likewise the terms used to explain the machine metaphor, which range

27 E[ugène] E[mmanuel] Viollet-Le-Duc, *Discourses on Architecture* (Boston: James R. Osgood, 1875 [1863]), p. 186. Cf. on this Jacques Lucan, *Composition, non-composition: Architecture et théories, XIXe–XXe siècles* (Lausanne: Presses polytechniques et universitaires romandes, 2009), pp. 268–71.

28 Le Corbusier's proposition found an early formulation in 1911 in the description of the Parthenon as a "terrible machine," cf. Ivan Zaknić (ed.), *Journey to the East: Le Corbusier (Charles-Edouard Jeanneret)*. Trans. Ivan Zaknić (Cambridge: MIT Press, 1987), p. 212. Corbusier's first known use of the expression "machine à habiter" is in the article "Maisons en série," *L'Esprit Nouveau*, 13 (1921), pp. 1525–42; here p. 1538. It is further elaborated in *Vers une architecture* (1923). Shortly after Le Corbusier, Walter Gropius likewise took up the notion. Cf. the manuscript "Wohnmaschinen," February 6, 1922, Bauhaus-Archiv, Berlin, 19/694.

29 Machinic terms and references have been used to address aesthetic issues in architecture at least since the mid-nineteenth century. For an early example, see Horatio Greenough, "American Architecture" (1843), in Greenough, *Form and Function: Remarks on Art, Design, and Architecture* (Berkeley: University of California Press, 1947), pp. 51–68.

30 See Reyner Banham, "Machine Aesthetic," *The Architectural Review*, 117 (1955), pp. 225–8; Banham, *The Architecture of the Well-Tempered Environment* (London: The Architectural Press, 1969).

31 Nicholas Negroponte, "Toward a Theory of Architecture Machines," *Journal of Architectural Education*, 23, no. 2 (1969), pp. 9–12; Negroponte, "Aspects of Living in an Architecture Machine," in Nigel Cross (ed.), *Design Participation: Proceedings of the Design Research Society's Conference* (London: Academy Editions, 1972), pp. 63–7.

32 Cf. exemplarily Kazou Shinohara, "When Naked Space is Traversed," *The Japan Architect*, 228 (1976), pp. 64–72; Shinohara, "The New Machine: Absorbing Chaos," *Columbia Documents in Architecture and Theory*, 1 (1992), pp. 135–45; Félix Guattari, "Les machines architecturales de Shin Takamatsu," *Chimères*, 21 (1994), pp. 127–41. On relations between Guattari, Shinohara, and Takamatsu, see Simone Brott, *Architecture for a Free Subjectivity: Deleuze and Guattari at the Horizon of the Real* (Farnham: Ashgate, 2011).

from "laboratory apparatus" to "manufacture," from "factory" to "tool," or, later, from "vehicle" to "computer." At least three different general models can be distinguished here, namely the mechanically propelled, the thermodynamically powered, and the cybernetic machine: three models distinguished by their different types of motion or motor, namely the vectorial, the transformative, and the informational. [33] The field is rendered yet more complicated if one includes the organic analogy, the history of which has long been deeply entangled with that of the machine. It is only with the rise of vitalism in the mid-nineteenth century that the model of the organic develops its thoroughly own, non-mechanistic explanatory power and even then it often remains in close dialogue with technical schemes. [34] Essentially, these various models are not mutually exclusive. They can be regarded neither as a sequence nor a development curve, since they are able to co-exist. Le Corbusier simultaneously references the tool and industrial enterprises. For Negroponte, the "architecture machine" is simultaneously a design technology and an inhabitable structure. Shinohara takes "machine" to refer to constructions of parts as well as to the transversalities of space. One and the same analogy thus gives rise to different instances of comparison.

6 Whatever the machine's type of operability and semantic nuance (mechanical, aesthetic, climatic, electronic, etc.), the overriding characteristic of the architecture-as-machine concept is the fact that it describes certain kinds of performance. The word performance must be understood here in its original sense, where it refers to the realization and the execution of a task, and more specifically, to the results achieved, as well as in its current theoretical sense in the field of science and technology, where it refers to the competence, the capabilities, or the effectiveness inherent to persons as well as to things. [35] From this perspective, architecture is not merely an inert artifact but also, by its very materiality, influences the behavior of its users, and prescribes certain human activities. It is anthropomorphic, not only because it is man-made and substitutes human activities, but also, and above all, because it shapes the actions and at times also the bodies of its users. [36] Somewhere between comfort strategies, disciplinary measures, and instances of control, the user is written into the machine's program, be it as the typified body of the industrial age or the fragmented body of the information age. The "mechanization" of architecture therefore does not isolate the material aspects of the building project from human relations. [37] Inversely, man is part of the

33 Michel Serres, *Hermes IV: La distribution* (Paris: Edition de Minuit, 1977), pp. 43–62.

34 See Joseph Rykwert, "Organic and Mechanical," *Res*, 22 (1992), pp. 11–18; Georges Canguilhem, "Machine and Organism," in Canguilhem, *Knowledge of Life*. Trans. by Stefanos Geroulanos and Daniela Ginsburg (New York: Fordham University Press, 2008), pp. 75–97.

35 Historically, cf. *The Oxford English Dictionary*, vol. 11 (Oxford: Clarendon Press, 1989), p. 544. Theoretically, cf. for example Andrew Pickering, *The Mangle of Practice: Time, Agency, and Science* (Chicago: University of Chicago Press, 1995).

36 Cf. Jim Johnson [=Bruno Latour], "Mixing Humans and Nonhumans Together: The Sociology of a Door-Closer," *Social Problems*, 35, no. 3 (1988), pp. 298–310; here p. 303. "Machines are lieutenants; they hold the places and the roles delegated to them," Latour concludes. Ibid., pp. 308–9.

37 Cf. Lefaivre and Tzonis, "Mechanization of Architecture" (see note 21), pp. 139–40.

architecture/machine: without him it cannot operate. At stake, for this same reason, are not only questions of technization or modernization, but instances of explication.

Indeed, the machinic in architecture is distinguished not only by the fact that it follows a certain program, triggers a certain process, or fulfills a certain procedure but also by its simultaneous revelation of the various determinants with which architecture is confronted, and the scope thus engendered for their modification, improvement, rejection, or re-design. Thus Le Corbusier describes the house as a "machine for living" in order to turn the spotlight on the unseen potential of the space and construction economies; Reyner Banham sees in the technical accomplishments of the modern "curtain wall façade" a chance to break down the classic load-bearing, protective, and insulating wall into its individual, independent, and functional components; and Peter Sloterdijk describes air-conditioning with its heating, cooling, humidification, and cleaning procedures as "an explication of the properties of air." [38] Accordingly, that which characterizes the technical device is not so much the technology underpinning its operation but rather the associated revelation of latencies and concealed procedures. [39] For here—and it is this which characterizes the machinic—"technical activity [becomes] a conscious operation that enters into a regulated relationship with the sciences [and] gains in coherence [whereby] technology becomes a theoretical issue." [40]

7 Architecture machines are hence neither isolated nor autonomous. A further analogy is therefore called for. Lewis Mumford tried as early as 1934 in *Technics and Civilization,* and moreover in his later work *Myth of the Machine,* to define the cultural and social role of the machine and to distinguish between "machines" in the sense of concrete technical objects (among which he implicitly ranked architecture too) and "the machine," by which he meant a society's entire technical complex: in his view, a prerequisite of the former. Mumford stresses that this is no idle wordplay, but rather that a classic definition—for example, Franz Reuleaux's description of a machine as a combination of parts, specialized in function, operating under human control, and utilizing energy to perform a task—serves equally to characterize social constellations. [41] This thesis was to have a profound and intricate influence on philosophical thought, particularly in France: when Deleuze and Guattari analyze the capitalist social order and the modern state as a "megamachine", or Michel Foucault famously goes on to describe disciplinary society through the spatial concept of the "panoptic machine," [42] their indebtedness to it is patent.

38 Peter Sloterdijk, "Airquakes," *Environment and Planning D: Society and Space,* 27 (2009), pp. 41–57; Sloterdijk, *Schäume,* vol. 3: *Sphären: Plurale Sphärologie* (Frankfurt: Suhrkamp, 2004), pp. 501–67.

39 Martin Heidegger, "The Question Concerning Technology," in Heidegger, *The Question Concerning Technology and Other Essays.* Trans. William Lovitt (New York: Garland Publishing, 1977), pp. 3–35; here pp. 16–17. Heidegger, however, was no friend of the concept of architecture-as-machine: "[It] is only a further sign of the prevailing groundlessness of thought and understanding today," he posited in 1929/30, "when we are asked to regard the house as a machine for living and the chair as a machine for sitting." Martin Heidegger, *The Fundamental Concepts of Metaphysics: World, Finitude, Solitude.* Trans. William McNeill and Nicholas Walker (Bloomington: Indiana University Press, 1995), pp. 215–16.

40 Erich Hörl, "Die offene Maschine: Heidegger, Günther und Simondon und die technologische Bedingung," *MLN,* 123 (2008), pp. 632–55; here pp. 639–40 (Trans. by the authors).

41 Lewis Mumford, *Technics and Civilization* (London: Routledge, 1934), pp. 12, 41–5; Mumford, *The Myth of the Machine,* vol. 1: *Technics and Human Development* (London: Secker & Warburg, 1967), pp. 188–94.

42 Gilles Deleuze and Félix Guattari, *Anti-Oedipus.* Trans. Robert Hurley, Mark Seem, and Helen R. Lane (Minneapolis: University of Minnesota Press, 2000 [1972]), pp. 141–2; Deleuze and Guattari, *A Thousand Plateaus.* Trans. Brian Massumi (Minneapolis: University of Minnesota Press, 2005 [1980]), p. 428; Michel Foucault, *Discipline and Punish: The Birth of the Prison.* Trans. Alan Sheridan (New York: Vintage Books, 1995 [1975]), pp. 217, 309.

Deleuze again, reviewing *Discipline and Punish,* distinguishes even more clearly between concrete machines, such as the "machine-école," the "machine-hôpital," or the "machine-prison," and abstract machines; between the building with its concrete appurtenances, such as cells or rooms, and the diagram as an expression of the relations of force within the apparatus of power. The abstract machine is a machine inasmuch as it segments society in time and space: in the disciplinary society, on different levels; in the control society, by modulating flows. [43] Here, the signs are once again reversed. For it is generally supposed that the machine concerns a specific aspect of architecture, one that can be isolated and described, and to which certain functions might be assigned. Yet it could equally be argued that "the machine" or the "abstract machine" determines the preconditions. For the machine, per Deleuze, is social before it is technological or, as in the present case, architectural. [44]

Concrete machines, machine models and analogies, machinic explications, and abstract machines—these are some of the key terms of reference by which the following essays proceed: in a historical framework that spans over 250 years; in an architectural framework, because our concern is the built artifact; and, not least, in a socio-political framework, because it is this which shapes preconditions for the former. The essays pursue various approaches to the machinic in architecture, in a sequence that traces a dual trajectory, thematic and chronological: thematic, inasmuch as we endeavor to address in this compilation a broad and diverse array of programs, performances, and processes; chronological, inasmuch as we presuppose them all to have, even if not necessarily a causal then a temporal relationship to one another.

43 Gilles Deleuze, "Postscript on the Societies of Control," *October*, 59 (1992), pp. 3–7. On the machine concepts of Deleuze and Guattari, see Henning Schmidgen, *Das Unbewusste der Maschinen: Konzeptionen des Psychischen bei Guattari, Deleuze und Lacan* (Munich: Fink, 1997); Gerald Raunig, *Tausend Maschinen: Eine kleine Philosophie der Maschine als sozialer Bewegung* (Vienna: Turia + Kant, 2008).

44 Gilles Deleuze, *Foucault* (Minneapolis: University of Minnesota Press, 2009 [1986]), p. 39.

From the gta Archive
Compiled and commentated by Daniel Weiss

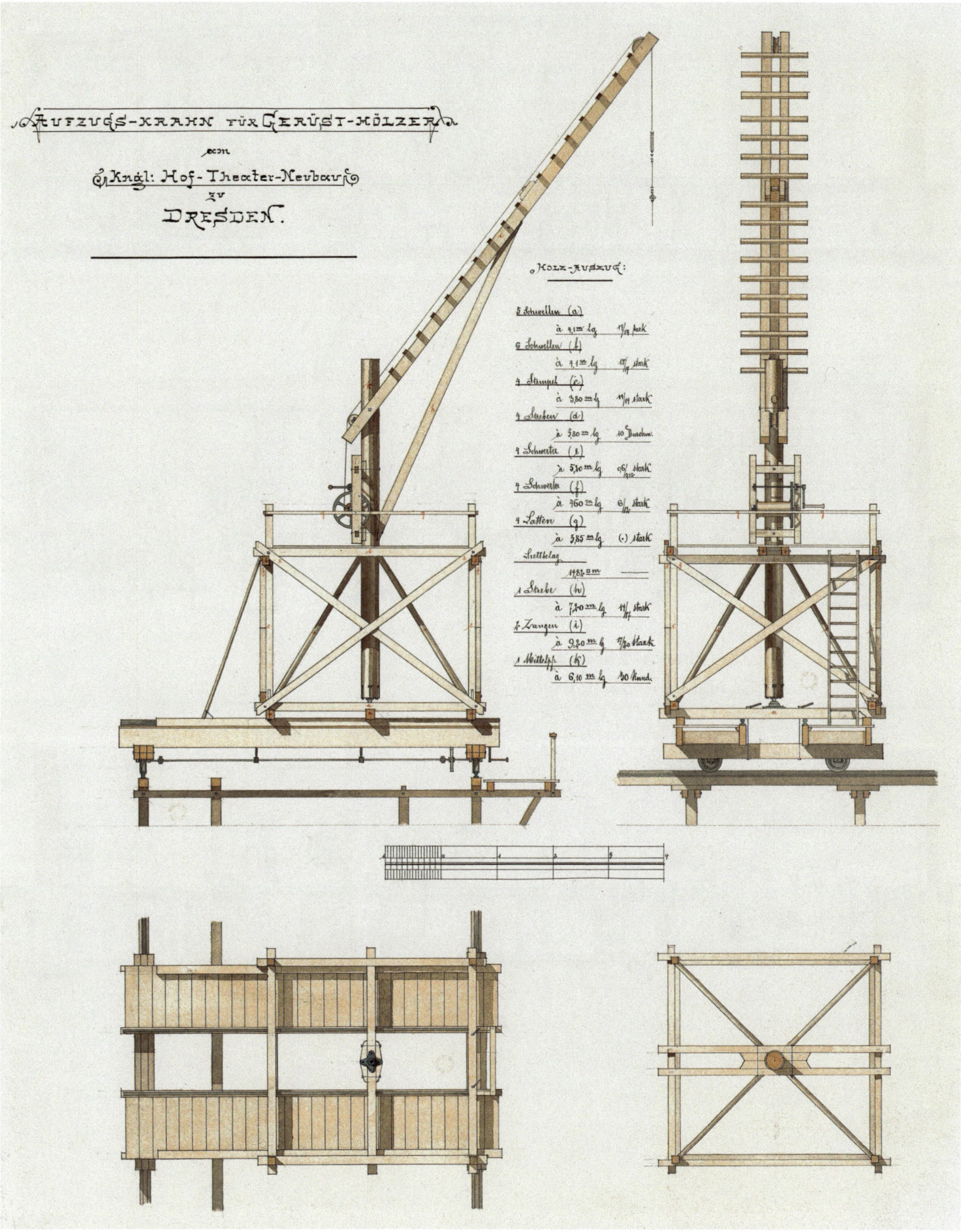

1 The architect's office of Gottfried Semper designed not only the structure and appointments of the new Royal Court Theater in Dresden. Accurate scale drawings in fine watercolors and Indian ink attest concern also with the construction of the scaffolding and the machines required to install it.

Gottfried Semper architect's office, Royal Court Theater in Dresden; ground plan and view of the lifting crane for the scaffolding's timber beams, c. 1871; Bequest of Gottfried Semper, gta Archives, ETH Zurich.

2 By the late nineteenth century, empirical research in the mechanical engineering field was booming at universities. The Federal Polytechnic School in Zurich consequently had the Semper student Benjamin Recordon build a spacious machine laboratory with an adjacent teaching building. As the elaborate plans show, its distinguished brick façade concealed state-of-the-art equipment for civil and mechanical engineering.

Benjamin Recordon, machine laboratory at the Federal Polytechnic School, Zurich; section through the machine hall, c. 1898; Bequest of Otto Rudolf Salvisberg, gta Archives, ETH Zurich.

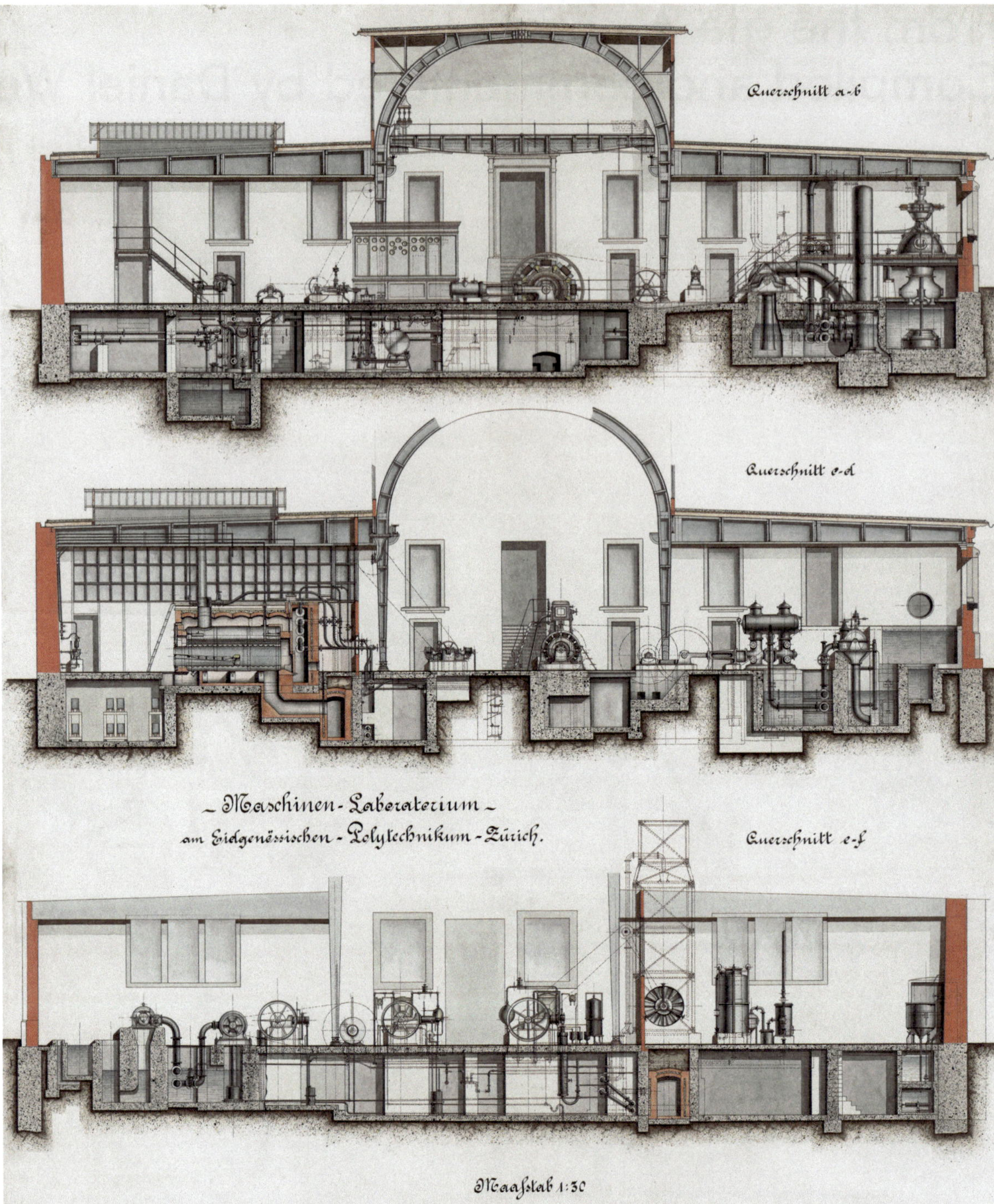

Warum

SIND UNSERE

MASCHINEN SCHÖN

WEIL SIE

Arbeiten
Sich bewegen
Funktionieren ?

FABRIK
SILO
LOKOMOTIVE
LASTWAGEN
FLUGZEUG

Warum

SIND UNSERE

HAUSER NICHT SCHÖN

WEIL SIE

Nichts tun
Herumstehen
Representieren ?

VILLA
SCHULPALAST
GEISTESTEMPEL
BANKPALAST
EISENBAHNTEMPEL

Es wäre unproduktiver Zeitverlust, wenn man heute beweisen wollte, dass man nicht mit eigenem Blut und einer Gänsefeder zu schreiben braucht, wenn die Schreibmaschine existiert. Heute zu beweisen, das die Aufgabe jedes Schaffens, so auch der Kunst, nicht Darstellen, sondern Dastellen ist, ist ebenfalls unproduktiver Zeitverlust. (Merz)

Die nächste Nummer des ABC wird Artikel und Entwürfe für das Bauen von Wohnungen und Städten bringen.

ATELIER
LICHT BILDER
LISSITZKY

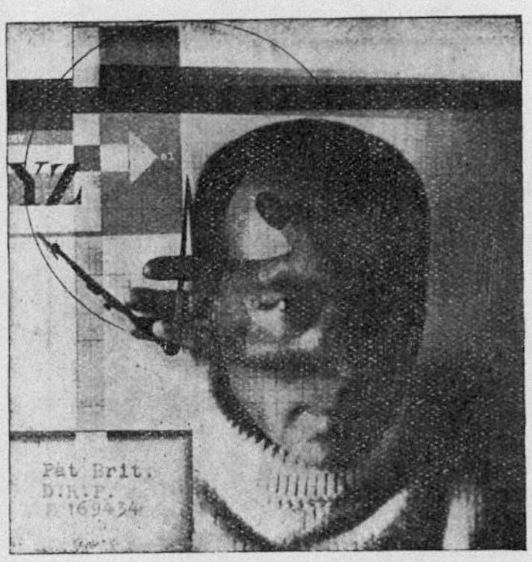

JEDER Künstler, der sich auf ABC abonniert, hat das Recht, seine Photographie gratis zu veröffentlichen. (Mit der Garantie, sofort berühmt zu werden). JEDE Dame, die sich auf ABC abonniert, hat das Recht, ihre Photographie gratis zu veröffentlichen. (ohne Garantie.)

JEDER abonniere sich auf ABC

NEUE BETONMATERIALIEN.

(Miteilungen aus der Praxis über technische und wirtschaftliche Eignung sind erwünscht)

GASBETON

Ziel: Ein leichtes, poröses Wandmaterial von geringer Wärmeleitzahl und genügender Festigkeit.

Herstellung: Gemisch von 40 Gewichtsteilen Zement und 60 Gewichtsteilen Schieferkalk, wird mit einem geringen Zusatz von Aluminium- oder Zinkpulver zur Erzeugung von Wasserstoffgas in Wasser angerührt und in flache Formen gegossen. Die Masse wird stark porös und kann nach dem Erstarren in Platten zersägt werden.

Resultat: Das Material lässt sich sägen, nageln, bohren, hobeln und verputzen. Porigkeit 76 %. Raumgewicht 700 kg/m³. Wärmeleitzahl 0,2 (Backsteinmauerwerk 0,65, Kiefernholz 0,15, Kork 0,05). Druckfestigkeit (nach 6 Wochen) 25-30 kg/cm². Zulässige Beanspruchung (nach Stockholmer Bauordnung) 3 kg/cm².

Anwendung: 1) Leichte Zwischenwände und Isolierungen an Stelle von Schlacken- oder Torfoleumplatten.
2) Isolierende, nicht belastete Zwischendecken mit der Möglichkeit leichter Armierung an Stelle von Bimskiesbeton.
3) Aussenmauern von geringem Gewicht und grösserer Wärmehaltung an Stelle der Backsteinmauer (die Stadt Stockholm lässt Mauern aus Gasbeton von 15-20 cm Dicke für 1-2stöckige Wohngebäude zu).

Hersteller: Axel Erikson, Architekt in Schweden.

STAHLBETON

Ziel: Schaffung einer möglichst harten, vor allem gegen mechanische Abnützung widerstandsfähigen Betonoberfläche.

Herstellung: Eine künstliche Eisenmasse, hergestellt durch Mischung von Portlandzement mit eigens präpariertem, metallischem Härtungsmaterial. Wird auf kaltem Wege in beliebiger Stärke aufgetragen.

Resultat: Das Material lässt sich ähnlich wie Gusseisen hobeln, drehen, bohren, feilen, sägen und schleifen. Infolge grosser Haftfestigkeit kann es als metallische Haut auf Beton, Naturstein, Bimsbeton Backstein, Leichtplatten, etc. aufgebracht werden. Härtegrad 2,2 mal so gross als bester Granit — infolgedessen weiteste Staubfreiheit. Rostsicherheit nur bei dauernder Benutzung. Zug- und Biegungsfestigkeit doppelt so gross, Druckfestigkeit 3-4 mal so gross als bei bestem Beton. Wasserdichtheit für eine Schicht von 20 mm bis auf 80 Atm. nachgewiesen.

Anwendung: Gebrauchsflächen von Böden, Randsteinen, Treppenstufen, Silos etc. Abdichtung von Wassergerinsen. Herstellung von Tresorwänden.

Hersteller: Patent von Prof. Dr. ing. Kleinlogel in Darmstadt.

TORCRETBETON

Ziel: Eine dünn konstruierte, widerstandsfähige, Wasser-, Witterungs- und Säurefeste äussere Schutzhaut für Bauten und Konstruktionsteile.

Herstellung: Auf eine maschenförmige Armierung von Rundeisen und 2-3 mm starken Drähten wird eine je nach den Anforderungen verschiedene Betonmischung (1 : 4 — 1 : 7) mittels Druckluft (Zementgun.) in 2 Lagen von ca. 30 mm Totalstärke aufgebracht.

Resultat: Die intensive Verbindung von Geflecht und Beton ergibt eine sehr elastische Haut bei geringster Gefahr von Schwindrissen. Durch Versuche wurde die Bruchgrenze (Zerreissen des Drahtes) bei Biegungsbeanspruchung mit 135 kg/cm² (0,5 % armiert) resp. 178 kg/cm² (0,65 % armiert) erreicht. Ein Torcretüberzug (von 5-20 mm) auf gewöhnlichem Beton blieb bei 5 Atm. Druck während 8 Stunden wasserundurchlässig.

Anwendung: 1) Dachhaut (bei beweglicher Auflagerung bis 11×40 m Oberfläche ohne Dilatationsfugen) auf U-Eisenpfetten in Abständen von 1 m aufliegend. Dicke der Haut: 30 mm, Mischung 1 : 4.
2) Ausbesserungen von schadhaften Betonkonstruktionen.
3) Dichten von Tunnels, Brückenuntersichten etc.
4) Schutzüberzug über Eisenkonstruktionen gegen Angriff von Feuer- und Rauchgasen.

Adressaten, **die nach Empfang dieser Nummer den Abonnementsbetrag nicht einzahlen (6 Nummern 4 Franken, Postcheck V 6971),) werden die folgenden Nummern nicht mehr erhalten. Sie werden gebeten, diese Nummer zurückgehen zu lassen.**

Abonnenten, **die No. 2 des ABC in beschädigtem Zustand empfangen haben, erhalten diese auf Antrage bei der Administration nachgeliefert.**

Probenummern werden auf Anfrage gerne zugeschickt.

3 With the launch of its journal in 1924, the avant-garde group ABC distanced itself from what it regarded as the formalist and monumental bourgeois-idealistic "Baukunst" (art of building). Manifesto-style texts and collaged images drew attention to the accomplishments of engineers, and celebrated the aesthetics and social benefits of modern technology with radically functionalist pathos. In the final issue, ultimately, in 1928, ABC propagated "The Dictatorship of the Machine."

"Warum sind unsere Maschinen schön?" in: *ABC Beiträge zum Bauen,* 3–4 (1925), n.p.

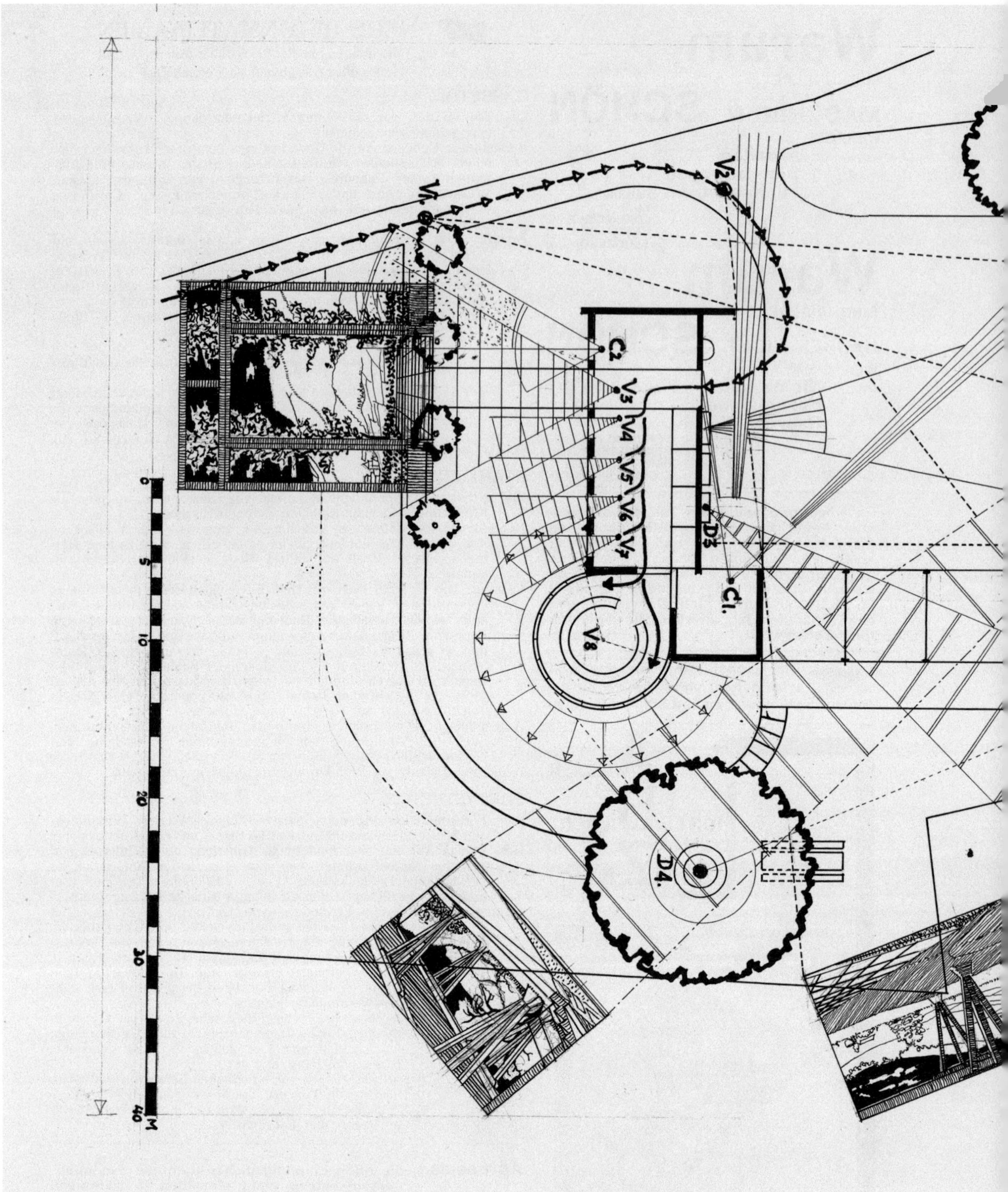

4 The cooperative children's home built in Mümliswil in 1939 was a chance for Hannes Meyer to show what he meant by scientifically founded organic functionalism. With the publication of two singular and at first glance seemingly abstract plans he turned the spotlight on the biomechanical and cinematographic interaction of mankind and architecture.

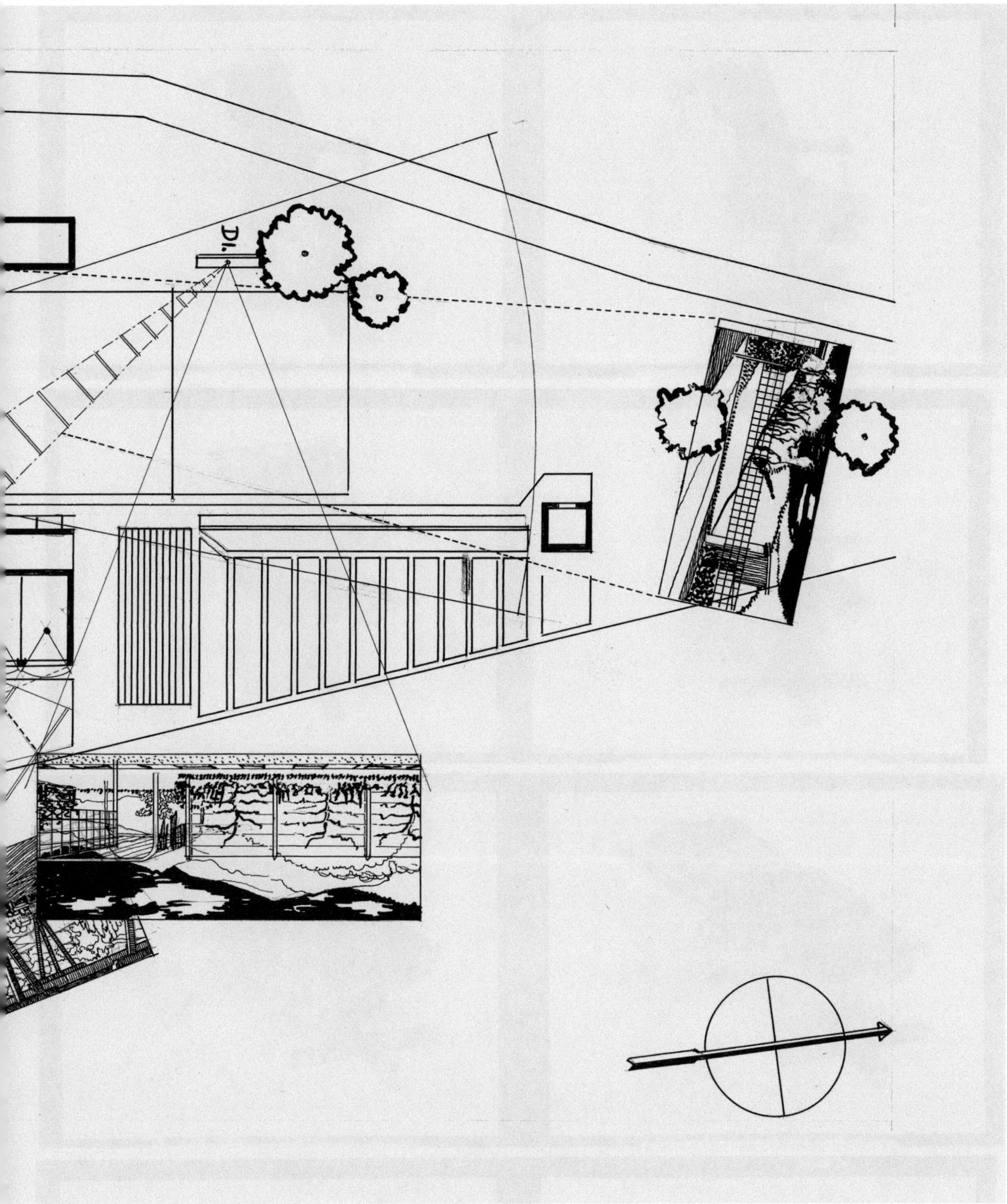

Hannes Meyer, cooperative children's home, Mümliswil; site plans with a schematic illustration of sight lines and circulation routes, c. 1939; Bequest of Hannes Meyer, gta Archives, ETH Zurich.

5 Among the comprehensive source materials compiled by the art historian Sigfriec Giedion in the course of research for his publication project *Mechanization Takes Command* there are numerous patent applications and severa patent furniture models from the latter half of the nineteenth century. Giedion acquired this "treasure trove of mostly unused ideas and lost lessons learned" at an auction of the United States Patent and Trademark Office in Washington, c. 1940.

Frederick H. Waste, Convertible Chair, U.S. Patent 169 752, 1875; estate of Sigfried Giedion, gta Archives, ETH Zurich.

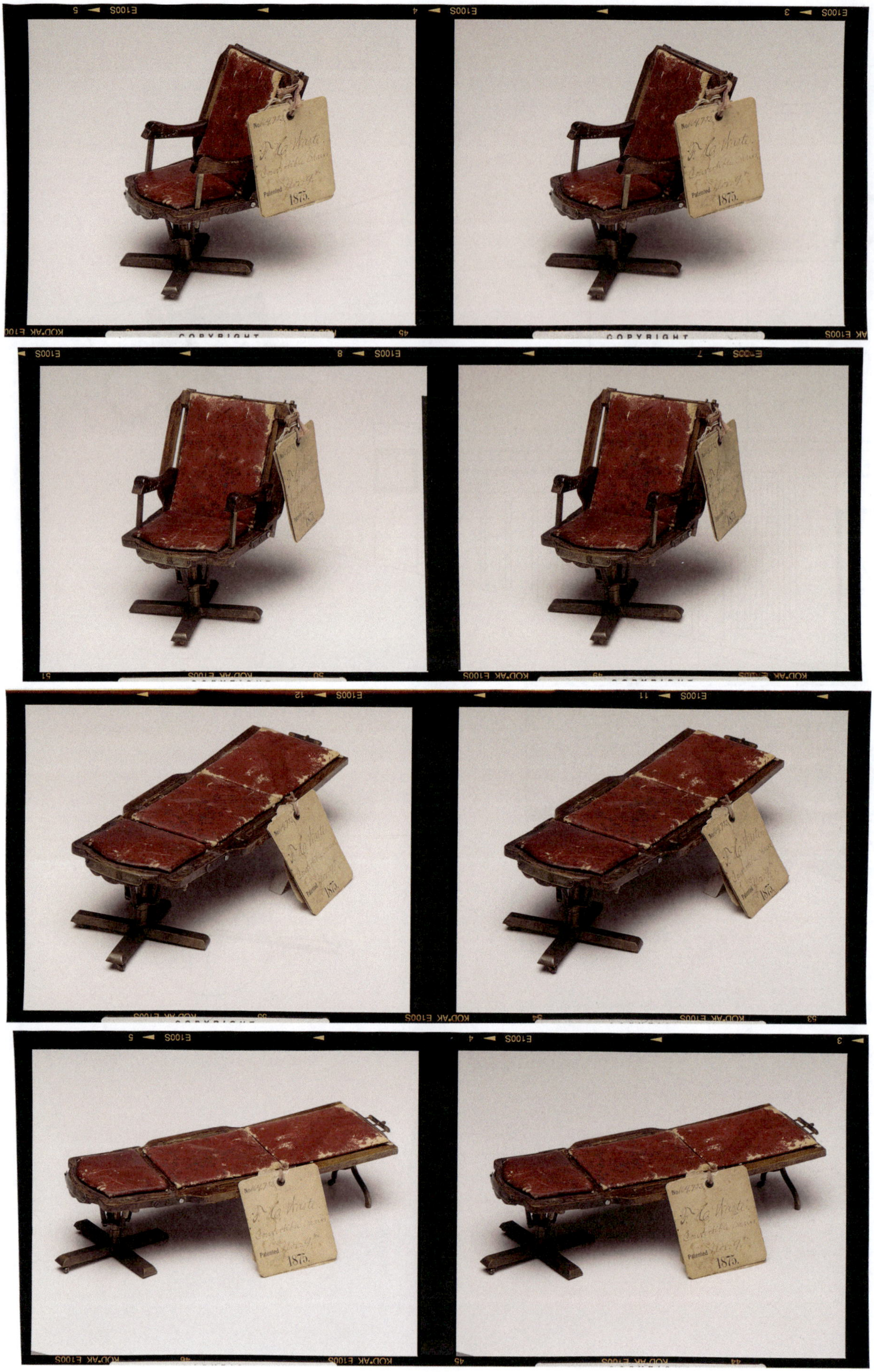

THE ANALYTIC APPROACH TO HOUSING

The old-time architect viewed the house as a collection of inviolable rooms, and his major problem was to arrange them conveniently and to enclose them in a shell that met the requirements of good taste. The analytical architect approaches the house as an engineer approaches the problems of designing a motor. He gets down to fundamentals. He assumes that all practices can be improved, is ruthless about discarding outworn ones. The drawing above depicts the basic considerations of the architect engineer: to provide shelter, privacy, heat, light, air, etc. With these in mind, he tackles the house not as a set of rooms but as a set of problems. One is to obtain privacy in one place, an outlook in another. The solution: windows of different sizes and shapes designed for specific jobs. Another problem is to admit as much sun as possible in winter and keep it out as much as possible in summer. The solution: large windows on the south side, so that the horizontal winter rays can enter, and a canopy or visor wide enough to block the vertical rays of summer sun.

to walk on barefoot. The bathtub would preheat itself automatically. The washbasin would be big enough to bathe the baby; like a kitchen sink it would be flush with a counter containing drawers and storage space. Soap would never turn to jelly because built-in soap dishes would drain properly. Space in the medicine cabinet would be sufficient to store the medicine for an army. And the mirror could be pulled close to a man's face and would actually be equipped with enough light to shave with ease.

The oft-made comparison of house and auto is open to some objection, but the analogy of bathroom and auto is accurate. The mass-production technique that made the auto what it is today may make the bathroom what it should be tomorrow.

Small bathrooms could be made in a few easy-to-assemble pieces, priced to undersell even the cheapest of home-installed jobs. The eminently sensible Victorian custom of a lavatory in each bedroom might be revived—and improved with devices like the disappearing washbasin of the Pullman bedroom. Large bathrooms could include an array of refinements and innovations. The largest might include dressing rooms equipped with everything from curling irons to overnight pants pressers. Such units could replace many an existing bathroom. With his twenty-year-old fixtures giving him trouble both esthetically and functionally, what house owner could resist the temptation to rip them out and put a $700 dream in their place? With only one bathroom for three large bedrooms, why wouldn't he install a $150 model in a big clothes closet, or a couple of small $50 lavatory units flush with the walls?

Designers have been asking questions and suggesting answers for several years. Six years ago, Buckminster Fuller got the Phelps-Dodge people (copper) to construct an experimental prefabricated bathroom of copper and aluminum. The war came along before the design could be developed in plastic, which Fuller expected to work better than metal. But Henry Kaiser, by whom Fuller is now retained, has announced his intention to get into prefabricated houses after the war—an intention that augurs well for the eventual development of a radically different bathroom. Just as much as new ideas, the bathroom needs engineer-hours, trials and errors, the willingness to take risks.

So does the kitchen, which for all the technical achievements

6 "Tomorrow's house" was to be understood not as an optimally composed sequence of spaces but as a nexus of elementary operations: the provision of shelter, warmth, privacy, calm, shadow, and exposure to light. The American architects George Nelson and Henry N. Wright tried to enlighten the general public in this regard by publishing various newspaper articles accompanied by lyrical didactic illustrations, then, in 1945, the compilation of their theses: *A Complete Guide for the Home-Builder.*

"The Analytic Approach to Housing", illustration in George Nelson and Henry N. Wright, "Housing for Human Beings: Give People Better Engineered Houses and Let Style Take Care of Itself," *Fortune Magazine*, April 1943, pp. 100–5, newspaper clipping from the source materials for *Mechanization Takes Command*; Bequest of Sigfried Giedion, gta Archives ETH Zurich.

7 When developing his thin concrete shells, the Swiss engineer Heinz Isler relied less on drawings or mathematical calculations than on practical experiment. He devised hundreds of models to test form-finding processes and examine the structural and static properties of shells. Pneumatic pressure and weights were used to exert force on miniature shells in static measurement models that were wired up to monitors by thick bundles of cable.

Heinz Isler, timber works Stäbler in Musberg; static measurement model of the domed shell, c. 1965; Bequest of Heinz Isler, gta Archives, ETH Zurich.

8 In several urban development studies he produced around 1970, Justus Dahinden, author of *Urban Structures for the Future*, addressed two modern social phenomena: mass consumption and the culture of leisure. The "socio-urban utopias" he proposed positively fused Pop culture dream worlds with exaggerated visions of a hi-tech future. Various combinations of synthetic mobile modular cells could be plugged into the external load-bearing structure of the leisure city Kiryat Ono while machinery within it, composed of podiums and crane systems, fostered a highly diverse range of public practices and events.

Justus Dahinden, the leisure city Kiryat Ono near Tel Aviv; exhibition panel showing a section of the Hill City, c. 1971; Bequest of Justus Dahinden, gta Archives, ETH Zurich.

9 The Variel system of prefab modules first developed by the Swiss architect Fritz Stucky in 1958 was subsequently used worldwide to build schools, hospitals, and office buildings, as well as apartment buildings and villas. Stucky took care of the entire process, not only design and production of the prefab elements, but logistics and distribution too. In the mid-1970s, Stucky parodied assembly line manufacturing in a comic-strip-style publication.

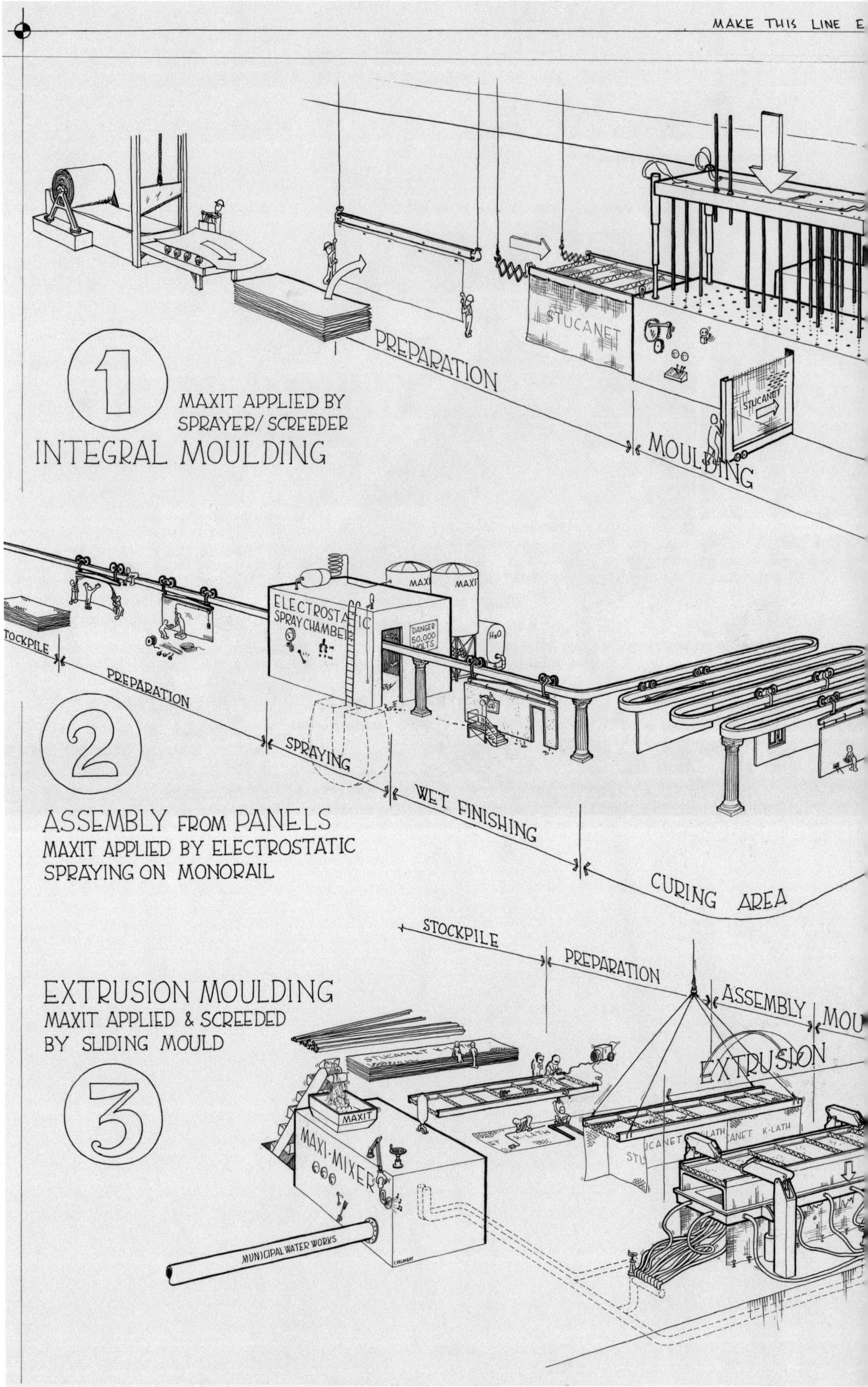

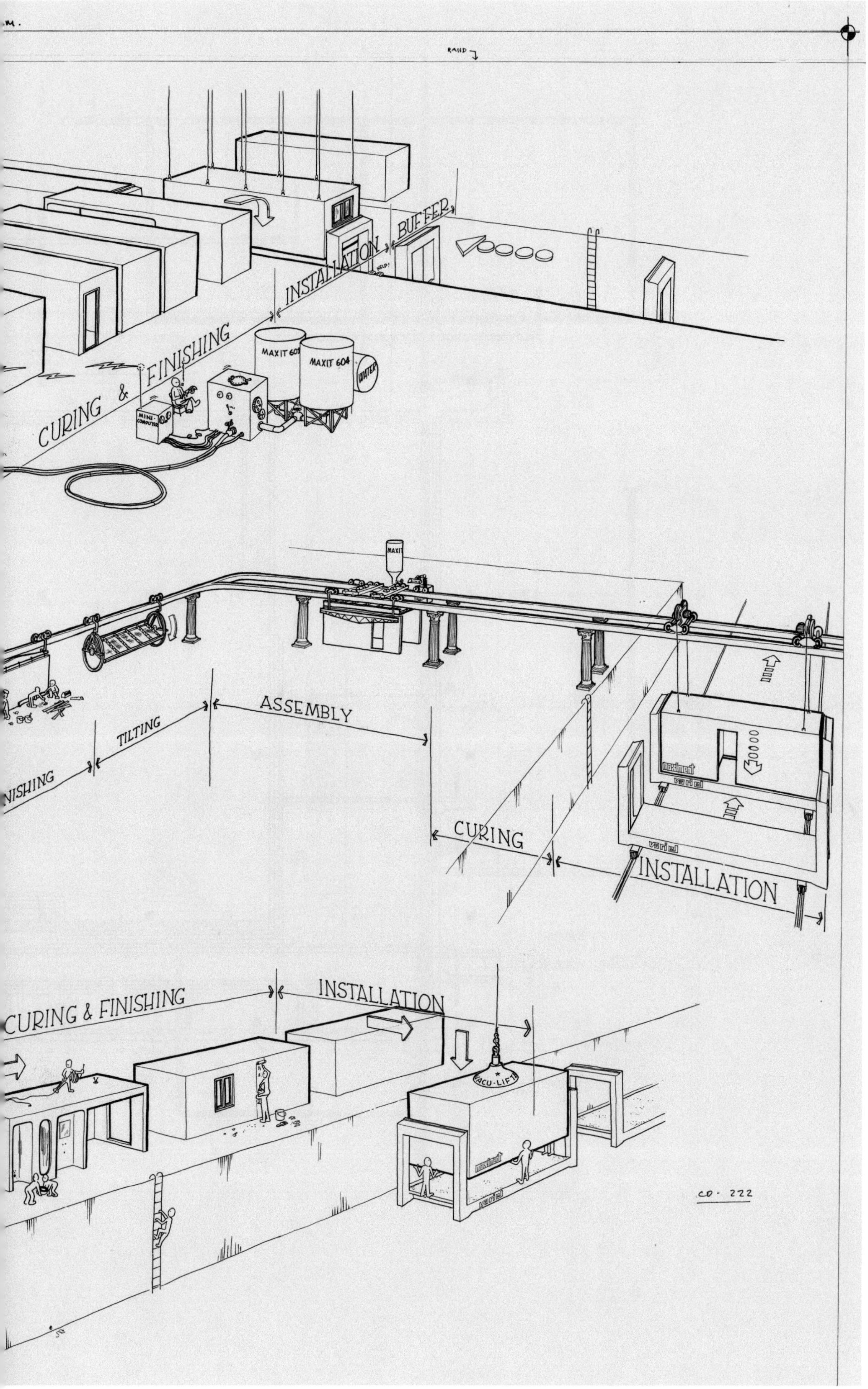

Fritz Stucky, production process for the Variel system of prefab modules, c. 1975; Bequest of Fritz Stucky, gta Archives, ETH Zurich.

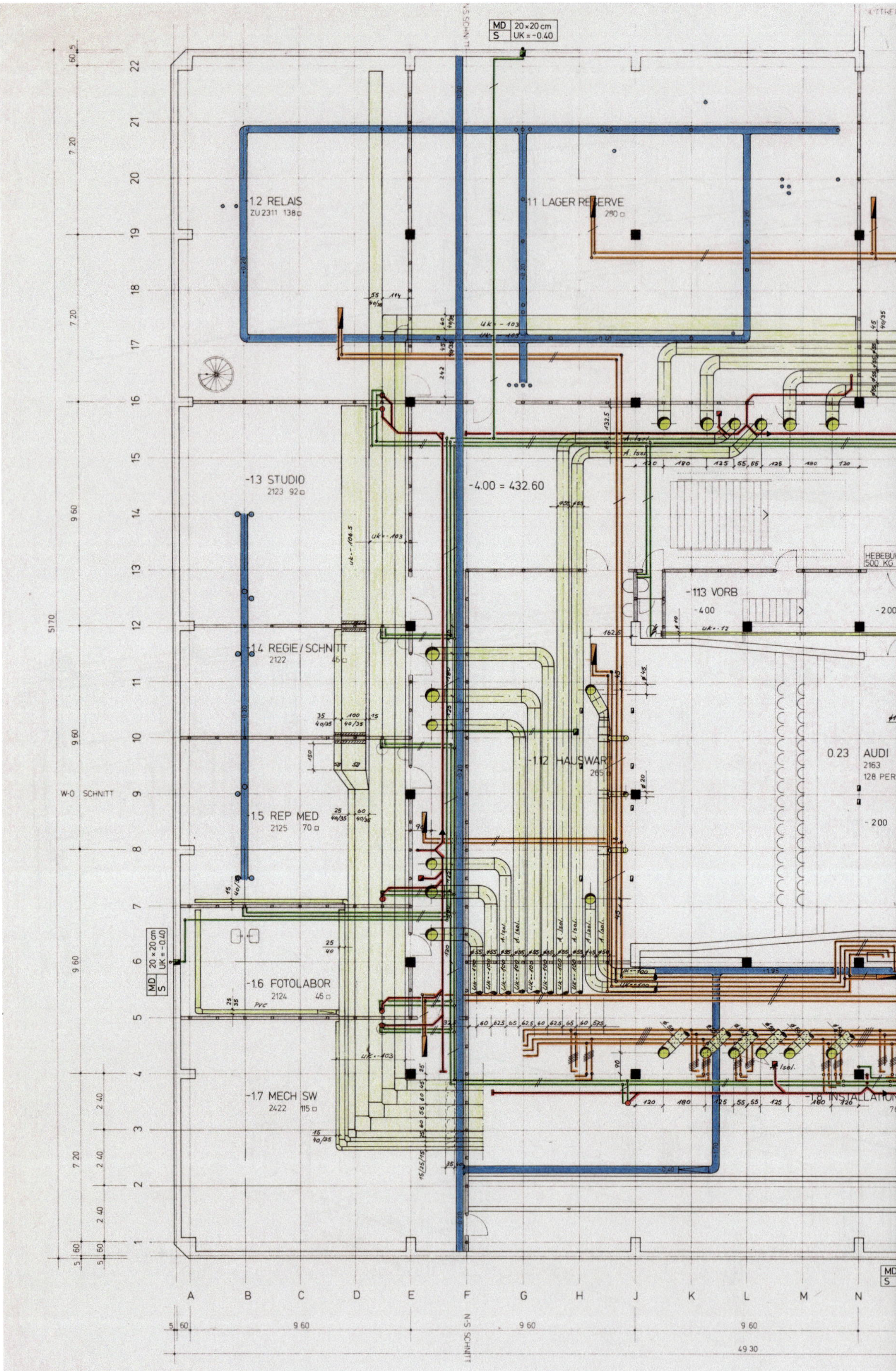

10 Is it possible to systematize all the building services in a well-equipped building as consistently as one can the construction of the latter and, above all, to coordinate the two? These are the questions to which Fritz Haller devoted his attention when planning the Swiss Federal Railways Training Center in Murten. In order both to install the various distribution networks independently of one another yet economically, and to leave leeway for future modifications, their arrangement was worked out in advance in a 3D planning grid.

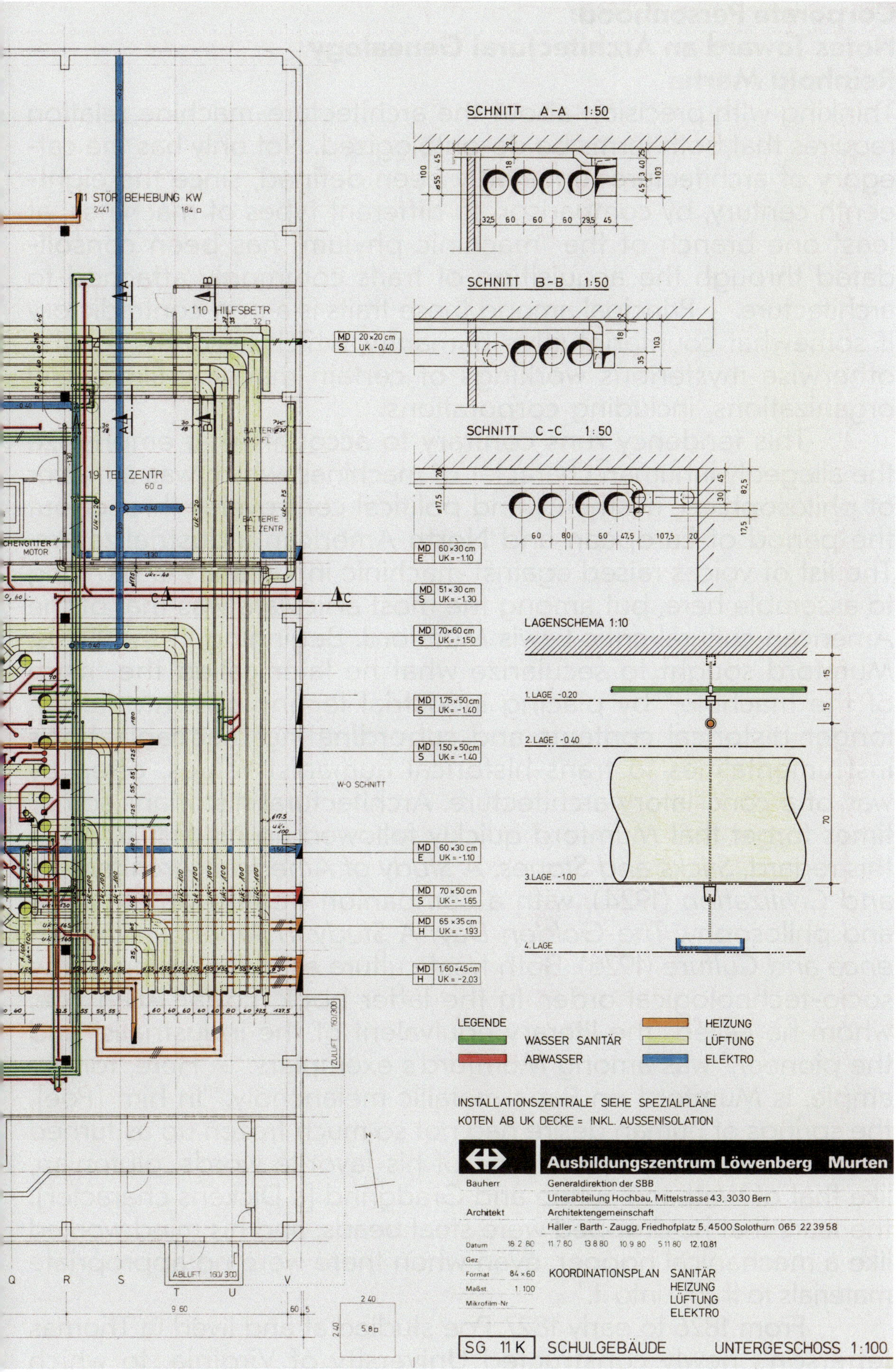

Fritz Haller, Alfons Barth, and Hans Zaugg, Swiss Federal Railways Training Center Löwenberg in Murten; coordination plan for technical installations, 1980; Bequest of Fritz Haller, gta Archives, ETH Zurich.

Corporate Personhood: Notes Toward an Architectural Genealogy

Reinhold Martin

Reinhold Martin is Professor of Architecture at the Graduate School of Architecture, Planning, and Preservation at Columbia University.

Note: I am grateful to Stefan Andriopoulos for his comments on an earlier draft of this text.

Thinking with precision about the architecture-machine relation requires that both terms be de-ontologized. Not only has the category of architecture repeatedly been defined, since the eighteenth century, by comparison to different types of machines; at least one branch of the "machinic phylum" has been consolidated through the acquisition of traits commonly attached to architecture. [1] Principal among these traits is a non-contradictory if somewhat counterintuitive humanity, which accounts for the otherwise mysterious workings of certain machine-like social organizations, including corporations.

This tendency runs contrary to accounts that emphasize the allegedly inhuman character of machines, which was a matter of philosophical, aesthetic, and political controversy throughout the period of European and North American industrialization. The list of voices raised against machinic inhumanity is too long to assemble here, but among the most articulate was that of the American cultural critic Lewis Mumford. Beginning in the 1920s, Mumford sought to secularize what he later called the "myth of the machine," by placing industrial technology into larger, longer historical contexts and subordinating mechanization's instrumentalities to trans-historical humanist values, often by way of a conciliatory architecture. Architectural historians sometimes forget that Mumford quickly followed up his first effort in this regard, *Sticks and Stones: A Study of American Architecture and Civilization* (1924), with a companion treatise on literature and philosophy, *The Golden Day: A Study in American Experience and Culture* (1926). Both treat culture as a dimension of the socio-technological order. In the latter book, Edgar Allan Poe, whom he called "the literary equivalent of the industrialist and the pioneer," was among Mumford's exemplars. [2] Here, for example, is Mumford on Poe's metallic melancholy: "In him [Poe], the springs of human desire had not so much frozen up as turned to metal: his world was, in one of his favorite words, plutonian, like that of Watt and Fulton and Gradgrind [a Dickens character]: the tears that he dropped were steel beads, and his mind worked like a mechanical hopper, even when there were no appropriate materials to throw into it." [3]

From 1826 to early 1827, Poe studied at and lived in Thomas Jefferson's newly constructed University of Virginia, to which we will eventually return. For now, it is enough to note that, in *Sticks and Stones*, Mumford celebrated Jefferson's neo-Palladian design for the Virginia campus as "perhaps [the] most perfect

1 On the protean flux of the "machinic phylum," see Gilles Deleuze and Félix Guattari, *A Thousand Plateaus: Capitalism and Schizophrenia*. Trans. Brian Massumi (Minneapolis: University of Minnesota Press, 1987), pp. 409–10.

2 Lewis Mumford, *The Golden Day: A Study in American Experience and Culture* (New York: Boni and Liveright, 1926), p. 76.

3 Ibid.

consummation" of the "classical order" in the early American republic, an order that, according to Mumford, rapidly disintegrated "under the combined influence of pioneer enterprise, mechanical invention, overseas commerce, and the almost religious cult of utilitarianism." [4] Writing a few pages later about the partnership of neo-Gothic romanticism and utilitarianism in the new instruments of industrialized disintegration, Mumford observed that:

"The erection of factories, the digging of canals, the location of furnaces, the building of roads, the devising of inventions, not merely exhausted a great part of the available capital; even more, it occupied the energy and imagination of the more vigorous spirits. Two generations before, Thomas Jefferson could lay out and develop the estate of Monticello; now, with many of Jefferson's capacities, Poe could only dream about the fantastic Domain of Arnheim. The society around Poe had no more use for an architectural imagination than the Puritans had for decorative images; the smoke of the factory chimney was incense, the scars on the landscape were as the lacerations of a saint, and the mere multiplication of gaunt sheds and barracks was a sign of progress." [5]

The "Domain of Arnheim" to which Mumford contrasts Jefferson's Monticello plantation refers to an eponymous short story in which Poe conjures a picturesque landscape garden engulfing "a mass of semi-Gothic, semi-Saracenic architecture" (Poe's words) built to gratify his wealthy protagonist's aesthetico-erotic dreams. [6] "Desire is real!" exclaimed Mumford, who associated earthly desires, honestly expressed – rather than illusory, alienated dreams of the sort he found in Poe – with a full-fledged humanity able to confront the despairing nullity of machines, and asked, "But if sexual desire, why not every human desire?" [7] Mumford failed to recognize, however, the many channels through which machines elicit desire, and in the process acquire indelibly human characteristics. He was therefore unable to shed the grounding distinction between the mechanized, routinized utilitarianism of finance capitalism and its infrastructures that had grown up around him (and before him, Poe), and the "golden day" of human, American "experience" manifest in the naturalist philosophies of Ralph Waldo Emerson and Henry David Thoreau, mapped in the literary cartographies of Herman Melville, and built in the rough-hewn stone castles of Henry Hobson Richardson.

Ranged against this minority alongside Poe's Arnheim was also, according to Mumford, the stylized eclecticism of John Haviland's "Tombs" prison in New York (1838), the nickname of

4 Lewis Mumford, *Sticks and Stones: A Study of American Architecture and Civilization* (New York: Boni and Liveright, 1924), p. 68.

5 Ibid., pp. 82–3.

6 Edgar Allan Poe, "The Domain of Arnheim" (1847), in Poe, *Selected Tales* (New York: Oxford University Press, 1998), pp. 296–310; here p. 309.

7 Mumford, *Golden Day* (see note 2), pp. 280–1.

which derives, Mumford explains, from "the Egyptian character of its façade." [8] The Tombs, to which Melville's intransigent scrivener Bartleby was ultimately dispatched, was the very epitome of the Benthamite utilitarianism by which Mumford felt overwhelmed. It was also an exemplary instance of what Michel Foucault referred to as the apparatus, or *dispositif*, of "carcerality." [9] As readers of Foucault well know, machines, when understood as components of larger socio-technical apparatuses, participate in what Foucault called "subjectification." Apparatuses, or networks of institutions, technologies, laws, practices, and knowledge shape and reshape human subjectivity. Put more instrumentally, apparatuses produce subjects, but not without remainder, or the interplay of subjectivities through desire and other "arts of living," as Foucault made clear in his later work. Traditionally, this is taken to mean that apparatuses produce human subjects, in the sense that they, and the "grids" of power-knowledge that they lay out, including the human sciences themselves, render the human knowable and manageable as a historical construction rather than as a biological or metaphysical a priori. In short, in Foucault's works, subjects and subjectivities—disciplined, mad, visible, and so on—are recognizably human. The apparatuses, and the "human sciences" to which they are attached, only render their presupposed humanity contingent rather than absolute.

But subjects also produce apparatuses. What happens, then, when the impersonal apparatuses themselves begin acquiring human attributes? Does this not trouble the types of human-machine entanglements that Foucault so brilliantly analyzed? We are not speaking about mere personification, or about biological analogies common in the eighteenth century that persist to today. Nor do I refer to latter-day cyborgs, or cybernetic organisms. I am speaking instead about performative attributes. Attributes that certain systems or apparatuses began to acquire during the opening decades of the nineteenth century, at exactly that moment when Foucault situates the various "births"—of the carceral, of the clinic, of a certain type of madness, and of a certain way of ordering the world—by way of which a new type of "human" entered the historical, if mainly European, stage. [10]

More specifically, the becoming-human of modern apparatuses took the form of a "personhood" attained by those apparatuses, or by those elements of apparatuses, that had been known for centuries under the British crown as corporations. By 1800, especially in the newly independent United States, corporations were regularly formed to enable collective action like building roads or establishing cities semi-independently from the national state,

8 Mumford, *Sticks and Stones* (see note 4), p. 90.

9 Michel Foucault, *Discipline and Punish: The Birth of the Prison*. Trans. Alan Sheridan (New York: Vintage Books, 1995), esp. part IV, chap. 3, "The Carceral," pp. 293–308. On Haviland's prison designs, see Norman B. Johnston, "John Haviland, Jailor to the World," *Journal of the Society of Architectural Historians*, 23, no. 2 (1964), pp. 101–5.

10 My reassessment of the Foucauldian legacy differs slightly from that advanced by recent interpreters, such as Giorgio Agamben or Roberto Esposito, who are also concerned with what Esposito has called the "*dispositif* of the person." See esp. Giorgio Agamben, "What Is an Apparatus?" in Agamben, *What Is an Apparatus and Other Essays*. Trans. David Kishik and Stefan Pedatella (Stanford: Stanford University Press, 2009), pp. 1–24; and Roberto Esposito, "The Dispositif of the Person," *Law, Culture, and the Humanities*, 8, no. 1 (2012), pp. 17–30.

which was viewed by many with suspicion. Hence, the decades immediately following U.S. independence saw the proliferating incorporation of towns, turnpike authorities, bridge companies, religious associations, colleges, schools, and many other institutions. [11] During the long nineteenth century, these corporations shifted under the law from being conceived as mere vehicles for collective activity, to being recognized as active agents with rights and responsibilities of their own. The basis of this agency is the "legal fiction" known as corporate personhood. Today, among such legal persons is the research university; but the strange logic of corporate personhood may be most visible in the university's precursor, the residential college.

11 Pauline Maier, "The Revolutionary Origins of the American Corporation," *The William and Mary Quarterly*, 50, no. 1 (1993), pp. 53–8.

In 1819, in a case known as "Trustees of Dartmouth College v. Woodward," the UnitedStates Supreme Court ruled that privately chartered institutions held contract rights comparable to those of private persons. Dartmouth College had been incorporated in 1769 (that is, prior to U.S. independence) by means of a charter granted by Britain's King George III, as was typical at the time. [12] In 1816, in order to resolve a long-running conflict between the College's ousted president, John Wheelock, and its board of trustees, the State of New Hampshire, where Dartmouth is located, sought to revise the College's charter in order to render its trustees answerable to state government. The Trustees objected, arguing that this violated the contract clause of the U.S. Constitution, which prevents the state from impairing "the Obligations of Contracts" among private individuals, or among individuals and the state. In 1819, the Court found that the charter amounted to such a contract, and hence the actions of the State were in violation of this constitutional clause. [13]

12 Ibid., pp. 56–7. On the history of the corporate charter, see Oscar Handlin and Mary F. Handlin, "Origins of the American Business Corporation," *The Journal of Economic History*, 5, no. 1 (1945), pp. 1–23.

13 For a detailed study of the Dartmouth case, see Francis N. Stites, *Private Interest and Public Gain: The Dartmouth College Case, 1819* (Amherst: University of Massachusetts Press, 1972). The "contracts clause" is to be found in Article I, Section 10 of the U.S. Constitution.

From there, it was but a few steps to the conclusion that privately held corporations were, in a legal sense, persons capable of entering into contracts and possessed of many if not all of the rights and obligations held by their biological or "natural" counterparts. Most of these steps were taken in the later nineteenth century, and in 1886, in "Santa Clara County v. Southern Pacific Railroad Co.," the Supreme Court ruled that corporations were entitled to equal protection under the law as provided to natural persons under the Fourteenth Amendment, which had been ratified in 1868 largely to secure equal treatment for freed slaves. This last fact is not incidental; even in the north, slavery was present on many college campuses and underpinned the antebellum political economy in which corporate personhood arose. [14] Finally, in 1910, in "Southern Railway Co. v. Greene," the Court concluded "[t]hat a corporation is a person, within the meaning of the Fourteenth Amendment,

14 On slavery in the early American colleges and universities, see Craig Steven Wilder, *Ebony and Ivy: Race, Slavery, and the Troubled History of America's Universities* (New York: Bloomsbury Press, 2013).

is no longer open to discussion." [15] And in 1926, no less a figure than John Dewey theorized "corporate personality" as, essentially, a concrete performative. Legal historians have supplied partial explanations as to how this came about, but most of these presuppose (unlike Dewey) an ontological distinction between natural and artificial persons that is abrogated by force of law. [16] Moreover, nearly all such accounts are purely discursive, giving little sense of how the corporate person was or is materially constituted. For this, architectural analysis offers some clues.

In his closing argument before the Supreme Court on behalf of Dartmouth College, the lawyer and orator Daniel Webster, who was a Dartmouth alumnus, exclaimed of his Alma Mater to the presiding justice, John Marshall, that it is "a small college. And yet *there are those who love it*." At which point Webster reportedly choked up, tears filling his eyes. [17] Strategically successful as it was, Webster's declaration of familial love for Dartmouth College was genuine. Not because his apparent spontaneity testified to true feeling rather than calculation—that, we cannot know—but because, as the Court's decision bore out, the abstraction called Dartmouth was well on its way to becoming worthy of a singularly human emotion. In other words, understood as a speech act, Webster's avowal of "love" for Dartmouth helped performatively to call its object into being as a corporate person.

f.1 Dartmouth Hall, Dartmouth College, 1784–1791. Photograph, 1865.

At the time Dartmouth was indeed small, consisting of about 95 students taught by a handful of faculty overseen by a president and a board of twelve trustees. [18] Its campus comprised a single building, Dartmouth Hall, which was an early example of the all-purpose, double-loaded phalanx- or phalanstery-like residential and educational hall typical of American colonial colleges. [19/f.1] It is unknown who designed the original Dartmouth Hall, which was constructed between 1784 and 1791,

15 "Southern Railway Co. v. Greene," 216 US 400 (1910). http://caselaw.findlaw.com/us-supreme-court/216/400.html (accessed March 8, 2016). For a summary of this history, see Joshua Barkan, *Corporate Sovereignty: Law and Government under Capitalism* (Minneapolis: University of Minnesota Press, 2013), chap. 3, "Personhood," pp. 65–86; and Martin J. Sklar, *The Corporate Reconstruction of American Capitalism, 1890–1916: The Market, The Law, and Politics* (New York: Cambridge University Press, 1988), pp. 49–53.

16 Barkan refuses this distinction, arguing instead that corporate personhood constitutes a *dispositif* or apparatus critical to "corporate sovereignty," which, like the *dispositif* of the "person" more generally, as theorized by the philosopher Roberto Esposito after Giorgio Agamben, operates a "ban" whereby the corporate entity is granted exceptional legal status or rights in the name, paradoxically, of fulfilling its societal obligations under the law. Barkan, *Corporate Sovereignty* (see note 15), pp. 76–86.

17 Kenneth E. Shewmaker (ed.), *Daniel Webster: "The Completest Man"* (Hanover, NH: Dartmouth College/University Press of New England, 1990), pp. 168–9. Emphasis in original.

18 Francis Lane Childs, "A Dartmouth History Lesson for Freshmen," *Dartmouth Alumni Magazine* (December 1957). http://www.dartmouth.edu/~library/rauner/dartmouth/dartmouth_history.html (accessed March 8, 2016). On the Dartmouth student body during Webster's time, which consisted of 150 students taught by a "handful" of faculty, see Robert V. Remini, *Daniel Webster: The Man and His Time* (New York: W.W. Norton, 1997), p. 44.

19 The College also made use of the top story of another building, since demolished, known as Rowley Hall, below which the building's owner ran a general store. John King Lord, *A History of Dartmouth College* (Concord, NH: Rumford Press, 1913), p. 122; and Bryant Franklin Tolles, *Architecture and Academe: College Buildings in New England before 1860* (Hanover, NH: University Press of New England, 2011), pp. 51–8.

burned in a fire in 1904, and was rebuilt in brick as an enlarged quasi-replica that stands today. [20] Webster's love for Dartmouth was likely consummated in that hall (if only in the Platonic sense) where he lived for three years, from 1797 to 1800, and where he and his 30 classmates performed regular recitations of classical verse. [21] That love would likely have been further secured in the after-hours antics in which he and his cohabitants no doubt indulged, as well as in his enthusiasm for public speaking, which on one occasion included a funeral oration for a classmate, although the young Webster was duly slighted by not being chosen to deliver the valedictory oration at the commencement ceremony. [22]

20 The entirely wooden original, which contained 16 dormitory rooms, plus a kitchen and a meeting room, and was completed in 1791, was most likely designed by Bezaleel Woodward, a college official, with the aid of drawings by carpenters Comfort Sever and William Gamble. Bryant F. Tolles Jr. and Carolyn K. Tolles, *New Hampshire Architecture: An Illustrated Guide* (Hanover, NH: New Hampshire Historical Society/University Press of New England, 1979), p. 287.

21 Remini, *Daniel Webster* (see note 18), p. 46.

22 Ibid., pp. 55–6.

Scattered accounts of college life in the early republic remind us not only of the relative youth of the exclusively male students like Webster, who was not from a family of great means,

f.2 Robert Smith and Dr. William Shippen, Nassau Hall (left), College of New Jersey (later Princeton University), 1757. Engraving by Henry Dawkins, 1764.

but also the relative lack of discipline that reigned over collegiate life. Probably the most infamous instance of indiscipline—which shows the disciplinary apparatus at its most functional rather than in dysfunction—is life at Princeton's Nassau Hall from about 1800 to 1820. f.2 Nassau Hall, a long, three-story phalanx-like building, was probably designed by Robert Smith, a Philadelphia builder, together with Dr. William Shippen, between 1755 and 1757, and was likely a model for Dartmouth Hall and many others. In 1802, Nassau Hall burned, leaving only the stone exterior walls. Immediately thereafter, the Hall was entirely rebuilt within these walls, with sturdier construction, a larger cupola, and small classicizing details, to designs by Benjamin Henry Latrobe. [23] Although it was

23 This version of the building lasted until another fire, in 1855, necessitated another total reconstruction within the original walls. On Latrobe's designs, see Paul Norton, "Benjamin Henry Latrobe's Nassau Hall," in Henry Lyttleton Savage (ed.), *Nassau Hall, 1756–1956* (Princeton: Princeton University Press, 1956), pp. 27–38.

not one of his most distinguished works, Latrobe's Nassau Hall elicited a number of incidents that exhibit what we can call the performativity of misbehavior.

Foucault suggests that delinquency is a product of the carceral system rather than its antithesis; failure is therefore among that system's prerequisites for proper functioning. In the sphere of education, a principal instrument for the proper distribution of failure is the examination, the inaugural instance of which is the entrance examination. Upon arriving in Hanover, New Hampshire, in 1797, the 15-year old Daniel Webster had his knowledge of English, Greek, Latin, and arithmetic tested before being allowed to enroll at Dartmouth. Such on-the-spot exams were common practice at the time. And so, during those years, young men coming from throughout the northeast and from parts of the south had to do the same upon arrival in Princeton, as they sought admission to what was still called (until 1896) the College of New Jersey.

In the early decades of the nineteenth century, the Princeton grounds — the country's first "campus" — consisted of five buildings: the President's House, a Professor's House, Geological Hall (now Stanhope, also designed by Latrobe), Philosophical Hall (also Latrobe, now demolished), and Nassau Hall. Originally, Nassau Hall comprised approximately 40 living chambers (some of which were used as recitation rooms), a prayer hall, library, and basement kitchen and dining room. [24] The edifice had virtually all of the attributes of Foucault's disciplinary apparatus: enclosure, or confinement; a system of cellular partitioning; distinctly marked "functional sites;" and "ranks," both within rooms (rows of beds, or desks) and among them (by year, etc.). [25] Likewise class schedules, daily recitations, the teaching of proper handwriting, with proper posture, the student-pen-paper-chair-desk interface, various prohibitions on time wasting, and so on. [26]

Foucault argues that when joined together into a disciplinary system, these properties combine to produce supple, trainable, "docile bodies." [27] But the bodies trained in Nassau Hall were hardly docile. On the contrary, the presidency of Ashbel Green in particular, which began in 1812 after a period of severely declining enrollments and what some considered a decline in piety, was punctuated by what Green called "every kind of insubordination." [28] During Green's first term, three students were expelled after gunpowder exploded in Nassau Hall; another was expelled for climbing the belfry and ringing the bell at three o'clock in the morning, while yet another broke into the Prayer Hall and vandalized a Bible by cutting a deck of playing cards into its leaves. The following year, 1813 to 1814, saw firecrackers set

24 Paul Norton, "Robert Smith's Nassau Hall and President's House," in Savage, *Nassau Hall* (see note 23), pp. 3–26; here p. 16.

25 Foucault, *Discipline and Punish* (see note 9), pp. 141–9.

26 Ibid., pp. 149–54.

27 Ibid., p. 155.

28 Thomas Jefferson Wertenbaker, *Princeton, 1746–1896* (Princeton: Princeton University Press, 1946), p. 155.

off in the Hall and graffiti scrawled on its walls. Then, on the night of January 9, 1814, in the words of one historian, "a cracker, consisting of a hollow log charged with two pounds of gunpowder, was set off behind the central door of Nassau Hall." Windows shattered, walls cracked, and a piece of the log crashed through the Prayer Hall door. [29]

29 Ibid., p. 156.

One of the two former students was brought to civil trial and other miscreants were later expelled. Still, further mayhem ensued, particularly in the building's long hallways before and after evening meals. One evening President Green, who referred in his memoir to the giant firecracker (or bomb) as an "infernal machine," performed the duly panoptic ritual of standing outside the refectory with a lit candle. He recalls: "They passed me in perfect silence and respect, but as soon as they had got out of sight" some "began the usual yell." [30] Exasperated, the faculty voted to dismiss two students. On April 6, Green wrote in his diary: "I took the examination of the senior class on belles lettres and wrote letters to the parents of the two dismissed students. The Faculty met in the evening and a pistol was fired at the door of one of the tutors." And then again, on April 7: "Attended examination. We had a cracker in the college today and in the evening a company of students in front of the campus behaved in a very improper manner." [31] If that was not enough, in 1817 students nailed shut all the building's entry doors, broke windows, and generally ran amok, shouting "Rebellion!" and "Fire!" [32]

30 Ibid., pp. 156–7.

31 Ibid., p. 157.

32 Ibid., p. 167.

The College of New Jersey was far from alone in its troubles. Dartmouth and Yale experienced similar disturbances, and in 1823, there were explosions in Harvard Yard, with 43 students expelled just prior to commencement. [33] A widely circulated 1828 report by the Yale Corporation and its faculty responded to the pervasive indiscipline, as well as to the devaluing of classical learning and declining religious piety to which many attributed it, by calling for the "*discipline* and *furniture* [or furnishing] of the mind," dedicated to "the art of fixing the attention, directing the train of thought." [34] Among the requirements for this discipline was "a substitute … for *parental superintendence*." That is, according to the Yale faculty, "The parental character of college government requires that the students should be so collected together, as to constitute one family; that the intercourse between them and their instructors may be frequent and familiar. This renders it necessary that suitable *buildings* be provided, for the residence of the students." [35] The buildings to which we have been referring provided the diagram: Nassau Hall and Dartmouth Hall, but also, at Yale, Connecticut

33 Ibid., p. 158.

34 *Reports on the Course of Instruction in Yale College* (New Haven: Hezekiah Howe, 1828), p. 7.

35 Ibid., p. 9.

Hall (1750), at Harvard, Massachusetts Hall (1718–1720), and at the College of Rhode Island (later Brown University), University Hall (1770), among many others. But my argument has also been that, although they fit neatly into the encompassing grids of Foucault's apparatuses, the acts of indiscipline, and of love, called forth in the halls of these institutions were among the conditions necessary for corporate personhood. Not only, as Foucault would have it, because disciplinary failure inscribed a vicious circle of subjectification for which docility and delinquency were two sides of a coin, but because the apparatus itself directly elicited familial as well as libidinal affect.

A final scene: In 1825 there was a masked, drunken, 14-person "riot" on the Lawn of Thomas Jefferson's recently opened — public but still incorporated — University of Virginia. 36/f.3 The following year, a 17-year-old Edgar Allan Poe enrolled at the university and took up residence on the Lawn, moving shortly thereafter to a room on the Western Range, in a section known as "Rowdy Row." 37 From his perfectly carceral cell, Poe witnessed fights (including the biting of an arm, which led to an expulsion), 38 gambled away what little he had, and read classics. f.4

Although not himself a troublemaker, Poe was indigent, and he withdrew after only a year, resentful of the wealthier, drunken classmates to whose company he was condemned. 39 A few years later, he eulogized, in the persona of Helen, "The beauty of fair Greece / And the grandeur of old Rome." 40 Some speculate that Poe's classical gaze in this ode to antique beauty was still fixed on Jefferson's neo-Palladian, though not exactly Roman, and certainly not Hellenic, campus architecture. 41 Regardless, we must remember that *its* gaze, like that of Bentham's panopticon, was fixed upon *him.* It is therefore tempting to ascribe the near madness that stalked Poe — and the hallucinatory explosiveness

36 Wertenbaker, *Princeton* (see note 28), p. 158.

37 Charles W. Kent, "Poe's Student Days at the University of Virginia," *Bookman,* 44, no. 5 (1917), pp. 517–25; here p. 520.

38 Edgar Allan Poe to John Allan, September 21, 1826. http://www.eapoe.org/works/letters/p2609210.htm (accessed March 8, 2016).

f.3 Thomas Jefferson, University of Virginia, Rotunda and Lawn, 1826. Engraving by Benjamin Tanner from Boye´s Map of Virginia, 1827.

39 Arthur Hobson Quinn, *Edgar Allan Poe: A Critical Biography* (Baltimore: Johns Hopkins University Press, 1998 [1941]), pp. 110–111. See also Kenneth Silverman, *Edgar A. Poe: Mournful and Never-Ending Remembrance* (New York: Harper Collins, 1991), pp. 29–36.

40 Edgar Allan Poe, "To Helen," in Poe, *Poems* (New York: Elam Bliss, 1831), p. 39. By 1841, Poe had changed these lines to read: "To the glory that was Greece / And the grandeur that was Rome." For a contextualization of the changes, see Quinn, *Edgar Allan Poe* (see note 39), pp. 177–9.

41 See, for example, Darlene Harbour Unrue, "Edgar Allan Poe: The Romantic as Classicist," *International Journal of the Classical Tradition*, 1, no. 4 (1995), pp. 112–9.

of his writings—to the implacable, inversely productive logic of the apparatus: a lifelong rage against the machine. But I want to risk instead another suggestion: that Poe's "Helen" is one name for the corporate person whose birth we have been witnessing, a figure shaped by the insubordinate love of those subjected to its iron will, even as it shaped them.

It is easy, then, to add the college or university campus to Foucault's list of "other spaces," or heterotopias, such as asylums, prisons, landscape gardens, and colonies, where social norms are both reproduced and inverted, as in a mirror. 42 More difficult is to diagram the economy of desire, love, and power that coursed through Poe's body, and that of his university, as he gazed upon the lineaments of "Helen," or, in only apparent contradistinction, as his hand later drew the textual lines of a picturesque, mytho-colonial, "neo-Saracenic" Arnheim. Or, for that matter, as his literary compatriot and successor Melville conjured the colonnaded (and again, mytho-colonial) "Egyptian character" of the old Tombs prison as he documented Bartleby's fate. 43 Mumford and many others could not have arrived at the problem of distinguishing human experience from machine experience if the two had not already been entangled. For, the confusion he and his poets may have felt was and remains an effect of machines having become embodied, soulful beings, beautiful and terrible, to be addressed with fear, love, and awe. From these beings, and not before them, the problem of distinguishing humans and their architecture from machines and their architecture was born.

f.4 Edgar Allan Poe Room, University of Virginia. Photograph by Rufus W. Holsinger, 1915.

42 Michel Foucault, "Different Spaces," in Foucault, *Aesthetics, Method, and Epistemology*. Ed. James D. Faubion. Trans. Robert Hurley et al. (New York: New Press, 1998), pp. 175–85.

43 Herman Melville, "Bartleby, the Scrivener: A Story of Wall-street," *Putnam's Monthly*, 2, no. 11 (1853), pp. 546–57, and 2, no. 12 (1853), pp. 609–15. Melville refers to the prison's "Egyptian character" in describing Bartleby's death there in the story's concluding lines, p. 614.

Water/Space: Ledoux and the *Disposition* of Architecture and Machines
Julian Jachmann

Julian Jachmann is Professor for the History of Art at the Universität Regensburg.

In its multifaceted diversity, the architectural and literary work of Claude-Nicolas Ledoux (1736–1806) can only be described as dazzling. The manifold topics provoked a plethora of heterogeneous appropriations by architects as well as art historians, who made the architect and pupil of Jacques-François Blondel (1705–1774) successively a revolutionary figure, a proto-modern autonomous architect, and a follower of physiocratic, encyclopedic, sensualistic, and freemasonic thoughts. [1] Apart from the wide scope of his work and his unpredictable fortune, the main reason for the characteristic ambivalence can be found in the only publication of Ledoux, *L'architecture considérée sous le rapport de l'art, des mœurs et de la législation*. The tome of monumental size and broad scope, published in 1804, comprises 125 copper-prints and 240 pages of text. In chapters of varying length, each usually discussing one building, the author oscillates between decade-old and recent projects, between fantastic ideas and realized buildings, between private, public, and technical structures. Moreover, the text is of a poetic, highly metaphorical character, merging several literary genres and styles. This literary "sfumato," as Michel Gallet aptly called it, is at least partly the result of the precarious political situation around 1800 – a time when Ledoux is not at liberty to speak

f.1 Graduation building at the Royal Saltworks at Arc-et-Senans, published by Claude-Nicolas Ledoux, 1804.

1 For a bibliography, cf. Julian Jachmann, "Claude-Nicolas Ledoux als Sensualist," *Zeitschrift für Kunstgeschichte*, 75, no. 3 (2012), pp. 351–72. For a helpful overview of recent approaches see Gérard Chouquer and Jean-Claude Daumas (eds.), *Autour de Ledoux: Architecture, ville et utopie* (Paris: Presses universitaires de Franche-Comté, 2008). For important interpretations see Daniel Rabreau, *Claude-Nicolas Ledoux (1736–1806): L'Architecture et les fastes du temps* (Bordeaux: W. Blake & Co., 2000); Anthony Vidler, *Claude-Nicolas Ledoux: Architecture and Social Reform at the End of the Ancien Régime* (Cambridge, MA: MIT Press, 1990). Regarding the systematic organization of production and of social structures, the utopian designs of Ledoux were compared with the works of Sade and Vaucanson in Jacques Perriault, "Le concept de machine et de système chez Ledoux, Sade et Vaucanson," *Culture technique*, 7 (1982), pp. 113–23.

freely about his success during the last years of the Ancien Régime. [2] Yet on closer examination the volume shows a surprising unity. Ledoux employs his partly realized project of the saltworks at Arc-et-Senans as the focal point of the book, expanding and amplifying this structure at least on paper to create an ideal city called "Chaux." Numerous buildings are integrated into an urban idea that derives from the topographical structure of an English landscape garden and the literary form of utopia. Creating and supporting an ideal community, the benevolent will of the architect replaces the divine creator. [3]

One of the main protagonists in his ideal city is without doubt the element of water. [4] Above all, water is the raison d'être of the saltworks at the heart of Chaux. In order to use spare brine from nearby saltworks, aqueducts lead from there to a new, convenient location near a forest, where the abundance of wood facilitates the operation of furnaces for salt production. Ledoux emphasizes the manifold roles of water in his realized buildings as well as in his publication. He presents a fantastic design for a bridge over the River Loue near the saltworks, pointing out its importance as a mode of transport, depicting a towpath and vessels. [5] The saltworks themselves welcome the visitor with an entrance hall designed as a grotto, which alludes to the sources of the saltwater. [6] And even the steam rising from the production buildings should—if we were to believe the architect himself—be employed in a way such as to identify or characterize the building as a location of salt production. [7] The energy of rivers and canals is put to use in mills. Furthermore, by constructing wells, horse ponds, and public baths, the architect makes use of water for public welfare. As Anthony Vidler puts it concisely: Ledoux basically feels "surrounded by water." [8] By controlling the force and employment of the universally important element of water, the architect himself acquires a universal role in the eyes of Ledoux.

Yet the author does not focus on technical buildings alone but presents a variety of private and public building types. And indeed, the responsibility for engineering tasks like the aqueduct, the waterworks, or the saltpans did not lie with him but with specialized engineers and local craftsmen. He even criticizes the form of the graduation building. f.1 In his opinion, the wooden supports are "d'une proportion svelte, d'une proportion gothique" and thus lack tectonic plausibility—at least in comparison to the Doric columns he prefers; and according to him the hipped roof violates the laws of art and nature in favor of simple forms. [9]

This criticism points us to a fundamental issue of architecture in the Early Modern period, which came to be more and

2 Michel Gallet, *Claude-Nicolas Ledoux: Leben und Werk des französischen "Revolutionsarchitekten"* (Stuttgart: Deutsche Verlags-Anstalt, 1983), pp. 7, 126–30.

3 Ledoux emphasizes repeatedly the universal role of architecture, e.g. Claude-Nicolas Ledoux, *L'Architecture considérée sous le rapport de l'art, des mœurs et de la législation* (Paris: C.N. Ledoux and H.L. Perroneau, 1804), pp. 8, 17, 28–9; at some points the term "machine" is employed, e.g. the "machine ronde" of the universe, pp. 55, 236.

4 The dominance over nature and especially water was a notion also prevalent in the Ancien Régime, e.g. in Versailles. Katharina Krause, *Die Maison de plaisance: Landhäuser in der Ile-de-France (1660–1730)* (Munich: Deutscher Kunstverlag, 1996), p. 33.

5 Ledoux himself stresses the importance of canals as a mode of transport, cf. Ledoux, *L'Architecture* (see note 3), p. 74.

6 Vidler, *Ledoux* (see note 1), p. 114.

7 Ledoux, *L'Architecture* (see note 3), p. 119.

8 Vidler, *Ledoux* (see note 1), p. 316.

9 Ledoux compares the structure to a ship or "ville flottante." See Ledoux, *L'Architecture* (see note 3), p. 57; Vidler, *Ledoux* (see note 1), p. 316.

f.2 Frontispiece of Bernard Forest de Bélidor, *Architecture hydraulique*, vol. 1, 1737.

more of a predicament in the eighteenth century. Architecture relied mainly on the Vitruvian ideas of stone and column buildings following classical examples. The tectonic principle of massive load-bearing supports, especially columns, is essential to this concept. **10** However, the construction of machines, which were supposed to control and augment movement and force, required a very different approach. Instead of a structure built in stone — or a wooden building painted or plastered to imitate a stone building — a wooden structure was most common within mechanics; parts that moved, wheels, tubes, ropes, and oblique levers forming complex structures for which weight and stability were just one determining factor among others. Bernard Forest de Bélidor (1698–1761) was one of the leading specialists in the most important fields of engineering — the construction of military and hydraulic machines. In his extensive and popular publication on hydraulics, first published from 1737 to 1753, two illustrations reveal a striking conflict between machines and the prevailing Early Modern concept of Vitruvian architecture. **11** An engraving on the first pages of the treatise shows an example of classical column architecture, encasing a complex apparatus, which is obviously responsible for generating the water pressure for the fountains shown in the background. In this case the machine is limited to an existence in the shadows of architecture. **f.2** In another illustration the conflict of different modes of design seems to be more balanced — with an outcome all the more disturbing. **f.3** Bélidor shows the pumping machine of La Samaritaine, built after 1712 on the site of an older predecessor in order to supply the royal palaces in Paris with water. In a manner that can only be described as absurd, this seemingly heavy and tectonic example of Vitruvian architecture rests on a much lighter and less solid base that is constructed from wood and conceived more

ARCHITECTURE HYDRAULIQUE

10 Since the time of Vitruvius, the Doric order is thought to derive historically from wooden structures. In the eighteenth century this concept found an ardent follower in Marc-Antoine Laugier (*Essai sur l'architecture*, 1753). While these theories resulted in important formal experiments, surprisingly they did not help bridge the gap between the system of the column orders and the practice of vernacular or technical wooden buildings.

11 Bernard Forest de Bélidor, *Architecture hydraulique, ou l'art de conduire, d'élever et de ménager les eaux pour les différens besoins de la vie*, part 1, vol. 1 (Paris: Charles-Antoine Jombert, 1737).

as a machine than as a building. 12 This obvious discrepancy between achievements in architecture and in engineering posed a problem in the eighteenth century—not least because of the growing professional specialization that manifested itself in successful engineering schools. 13

Bridging structural difficulties with "talking architecture"

If we follow Ledoux's lead in affirming the unifying and universal role of the architect, this problem had to be addressed somehow. Two examples from Ledoux's *L'Architecture*—a bridge and the Maison des Directeurs de la Loue—show how this problem could be dealt with on a conceptual level. The pathway to the ideal city of Chaux, which leads over a bridge across the River Loue, was most likely inspired by a similar project that Étienne-Louis Boullée (1728–1799) developed around 1787. 14/f.4/f.5 Ledoux introduces this

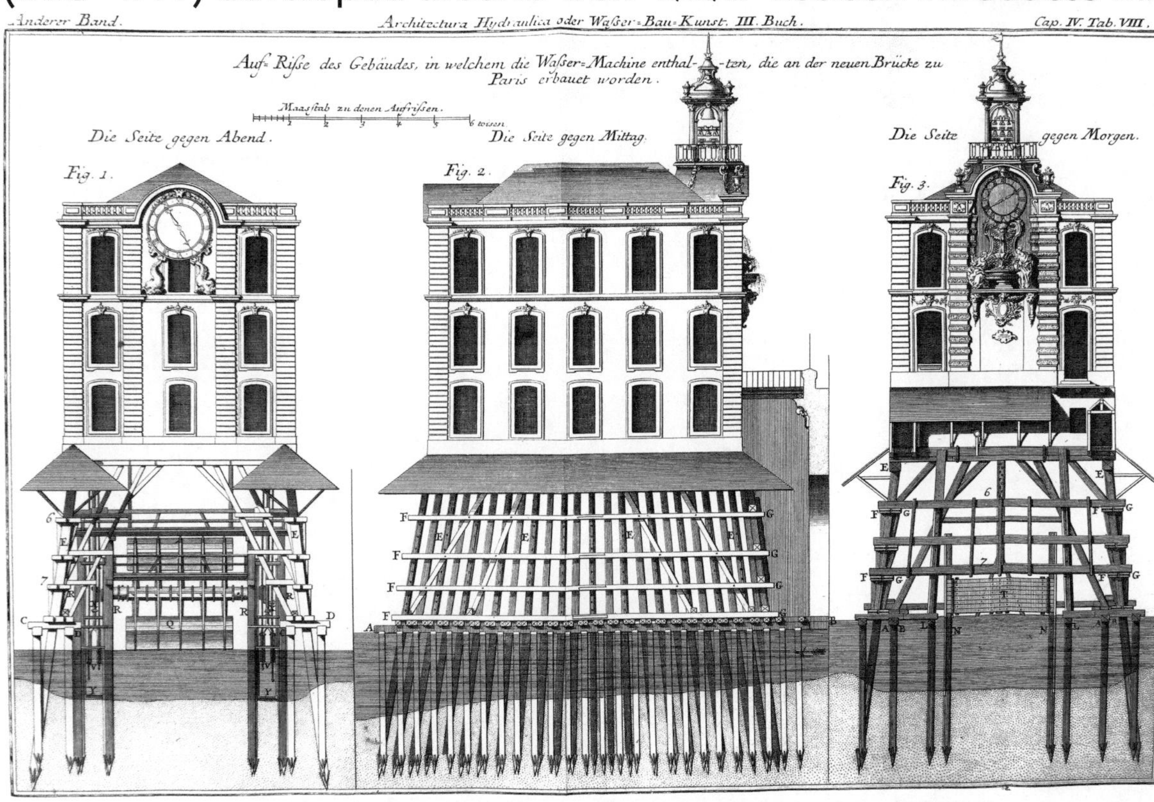

rather fantastic design with two illustrations. A perspective view shows the building in its context: a lavish landscape, interspersed with other imaginary building types, being part of his utopian concept. Furthermore, the function of the structure as junction of transportation routes becomes obvious as a towpath runs below street level. Ground plan and profile of the bridge reveal an implausible construction: the foundations are far too extensive and the form of the bulgy pillars does not follow any conventional reasoning. But even more striking is the outer appearance. The bridge is very low and bare of all ornament, except for the sculptural manifestation of antique ships that seem to carry the road. The piers have the form of hulls while lowered

12 The scaffold-like base of the building could hardly be seen while standing on the bridge; its design is therefore focused on this point of view.

13 The highly complex topic involves the definition of the academic disciplines, the history of institutions and pedagogy in France, and the relationship between art and science. For an introduction to this field of study cf. Antoine Picon, *Architectes et ingénieurs au siècle des lumières* (Marseilles: Parenthèses, 1988). Andrew Saint provides a succinct summary: "Architecture, a fellow-traveler clinging to the juggernaut of engineering's onslaught, never quite fell off." Andrew Saint, *Architect and Engineer: A Study in Sibling Rivalry* (New Haven, CT: Yale University Press, 2007), p. 33.

f.3 La Samaritaine in Paris, built after 1712, published by Bernard Forest de Bélidor, 1743 [1737].

14 Ledoux, *L'Architecture* (see note 3), p. 45. Boullée's project was aimed at remodeling the Pont Louis XVI (Pont de la Concorde) in Paris, designed by Jean-Rodolphe Perronet; Boullée explicitly criticized the aesthetic aspect of the engineer's concept. See Picon, *Architectes et ingénieurs* (see note 13), pp. 273–4. Boullée's drawing is depicted and described in Jean-Marie Pérouse de Montclos, *Étienne-Louis Boullée (1728–1799)* (Milan: Electa, 1997), pp. 76–8. His design differs significantly from that of Ledoux; Boullée's aims seem to be solely semantic while Ledoux combines semantic and structural ideas as the hulls and masts form the main part of the structure.

f.4 Bridge design by Claude-Nicolas Ledoux, published 1804. → 42/43

Le Doux Architecte du Roi.

VUE PERSPECTIVI

Bovinet Sculpsit.
T DE LA LOÜE.

masts replace the arches. The main argument of a semantically charged building becomes obvious enough in the accompanying text, in which Ledoux mentions the pontoon bridge that Xerxes used to span the Hellespont. This analogy goes as far as to place two of the ship-like pillars next to the shore, thus emphasizing the parallel while impeding the towing of the ships. By drawing on the famous tradition of ephemeral bridges, which have their roots in antiquity, the architect is able to design a structure that draws attention to its function as well as its historical connection to heroic deeds and technical feats.

In order to assess the strange morphing of tectonic structures into sculpture in an unbiased manner, we have to consider not only the contemporary tendency towards an *architecture parlante* but also the prevalent issues of bridge construction. In the second half of the eighteenth century Jean-Rodolphe Perronet

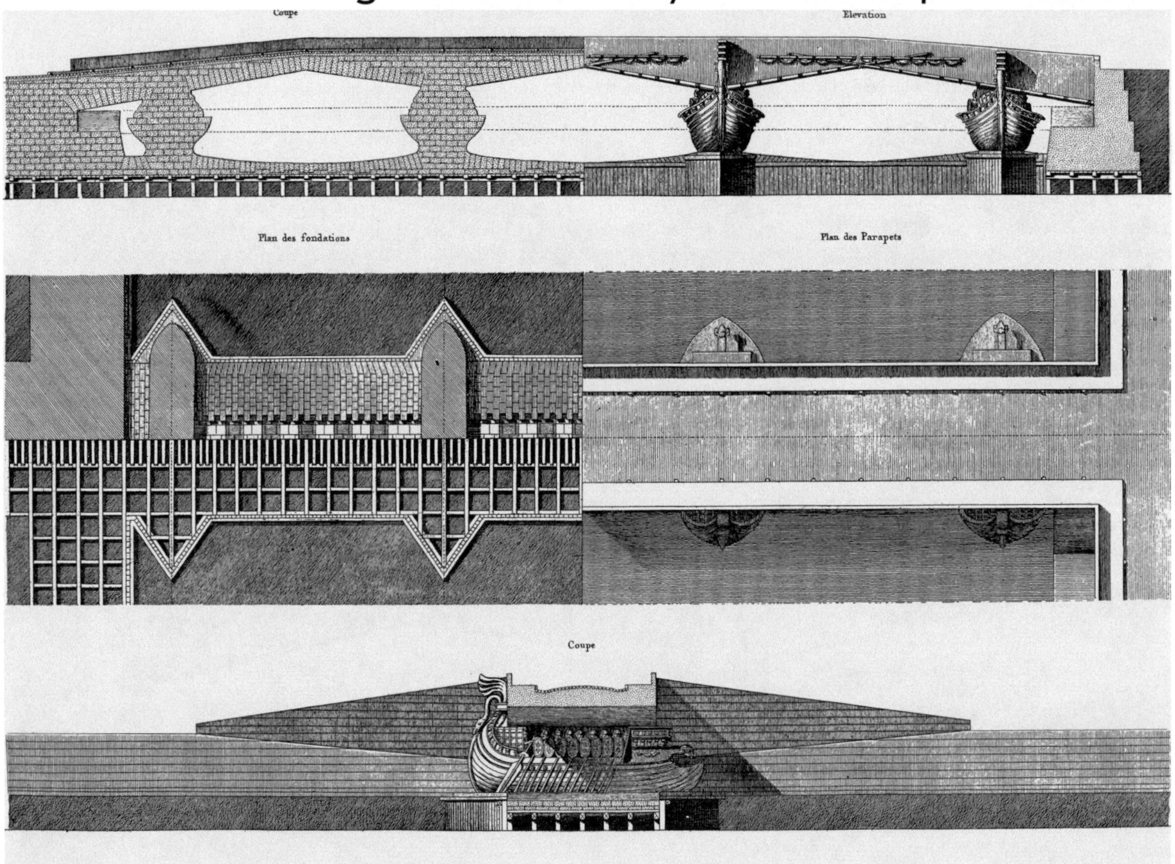

f.5 Bridge design by Claude-Nicolas Ledoux, published 1804.

(1708–1794) was by far the most prominent figure within this field of engineering. Not only did he realize several important and difficult projects, he also made these works accessible in an opulent publication, the *Description des projets*. [15] And what is more, Ledoux worked with him for some time. Among several other topics Perronet addresses the problem of pillars and arches. In order to reduce the contact surface of the bridge, which is subject to the constant attack of water flow and ice, he suggests the construction of smaller pillars of an adapted shape. Engineers of his time experimented with several alternatives,

15 Not all of the projects included in the *Descriptions* were designed by Perronet himself, e.g. the George V Bridge in Orléans (fig. 6) is a project by Jean Hupeau, in which Perronet was only involved at a very late stage. Cf. Jean-Rodolphe Perronet, *Description des projets et de la construction des ponts de Neuilly, de Mantes, d'Orléans & autres*, 2 vols. (Paris: Imprimerie Royale, 1782–1783).

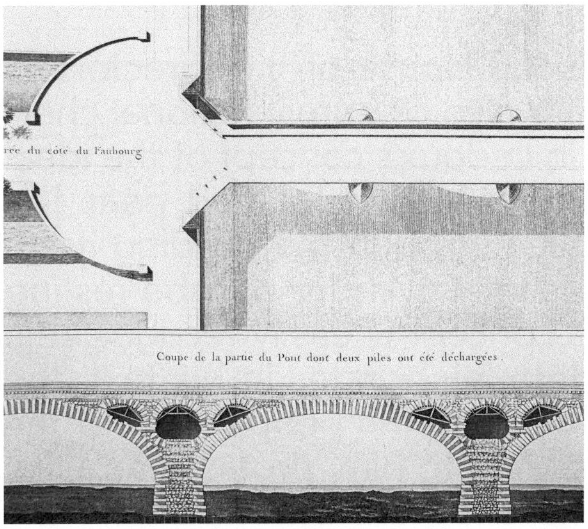

and sometimes ship-like forms were employed. **f.6** In addition, the whole structure could deviate from classical tectonic ideas. **16** The arch becomes shallow or stretched, constructed following a complex geometry. By chamfering the inner arch, he achieves a streamlined form—the famous "cornes de vache"—which served the purpose of reducing flood damage. And sometimes even the section through the pillars reveals a curved form that seems more dynamic than tectonic. In addition, the reduced piers as designed by Perronet were no longer able to absorb the extreme horizontal force exerted by the very low arches. The stabilization of the construction became the task of the abutments on the shoreline. **17**

Ledoux deals with this problem by disguising the building as a pontoon bridge, thus using shipbuilding and sculpture as a link between Vitruvian architecture and the demands of modern engineering. The thin arches as well as the curved and pointed pillars are thus ascribed to a historical paradigm. By choosing this solution, the author reveals a deeper understanding of the contemporary problems of bridge design. In light of the central issue of reducing the resistance to flow **18** the idea of using the forms already established in shipbuilding to design the pillars seems perfectly consistent. And the form of ships expresses very well the reduced structural role of the piers. Vessels are able to absorb large amounts of vertical forces—i.e. weight—while being horizontally unstable. Another problem is the lack of a universal method for constructing the arch, which Ledoux resolves in a rather bold manner by using straight lines, resembling the masts of ships. Thus he also introduces an element of timber architecture, as wooden bridges show similar V-shaped structures starting from the pillars. **19** This approach has certain implications regarding Ledoux's view of his profession. The architect is not only responsible for finding and exploiting corresponding traits and intersections between different disciplines but also for designing buildings that are appropriate to their context and communicate their meaning to the beholder.

f.6 George V Bridge in Orléans, designed by Jean Hupeau and built 1751–1763. Perronet published this building in 1783. The ship-like form of the pillars and their curved section are obvious, but Perronet's "cornes de vache" are still missing.

16 Perronet was well known for his very light constructions, and even today Picon calls his slender piers "spectaculaire," Picon, *Architectes et ingénieurs* (see note 13), p. 152. Picon also discusses Perronet and Ledoux. Cf. ibid., pp. 147–50, 273–5.

17 Ibid., pp. 147–51.

18 The engineer Emiland Gauthey (1732–1806) even gives a detailed depiction of resulting wave shapes caused by different pier forms, thus comparing the results in an almost diagrammatic way. See Emiland Gauthey, *Œuvres de M. Gauthey: Traité de la construction des ponts*, vol. 1 (Paris: Firmin Didot, 1809), plate XIII.

19 Ibid., vol. 2 (1813), plates II, III.

Hybrid: *maison–machine*

The second important example is the House of the Directors of the River Loue. 20/f.7 This building is also of purely fictional character but holds a crucial position in Ledoux's concept of the ideal city. We see an official's dwelling, which is located near the source of the river. The water passes through the building itself, which can be described as a horizontal cylinder or tube resting on a cubic base. Once again the concept is easily recognizable as *architecture parlante*. As Ledoux elaborates in his text, the inhabitants are responsible for protecting the people against the destructive powers of water. According to him, professionals like engineers, officials, or architects have to transform nature from a force of destruction into a productive power, a hostile wilderness into the cultivated realm of mankind. This is made abundantly clear by the perspective view. The landscape behind the building seems barely inhabitable; a jagged mountain rises abruptly behind a dramatic waterfall; trees are bent crooked by the force of the elements. The scenery in the foreground is completely different. It consists of idyllic meadows, which are pleasing to the eye of the beholder as well as economically productive for the miller, who is using the tamed force of the river. On closer inspection the cylinder itself proves to be a hybrid of two different technical models. In the first place it alludes to a faucet or outlet. Since Roman antiquity, the diameters of outlets have been used to measure the amount of water in order to distribute it evenly or to calculate the fees for the individual consumer. 21 Thus Ledoux highlights the duty and the right of the authorities to control water distribution. However, the cylinder is not completely open. Instead, a protruding room blocks the upper half. A mechanism to control the water flow by inserting a pane of wood or metal is not typical for a valve at this time. Instead, this mechanism belongs to the realm of reservoirs and channels, where it is known as a "knife gate," "sluice gate," or "lock." 22 It becomes clear that the main points of reference are flood control and the intention to produce an even flow of water for shipping.

In the seventeenth century important channels were built in France, and Voltaire would later call the Canal du Languedoc (since 1789: Canal du Midi) "le monument le plus glorieux" for Louis XIV. 23 As with his bridge design, Ledoux not merely chooses a semantically charged form but merges it with a functional structure, in this case the particular distribution of rooms. In the choice of his rooms and the structure of the floor plan he follows a conventional model that he himself varied in dozens of examples. It can be identified as a French *maison de campagne* as described

20 Ledoux, *L'Architecture* (see note 3), pp. 50–2.

21 Frontinus-Gesellschaft (ed.), *Wasserversorgung im antiken Rom* (Munich: R. Oldenbourg, 1982), pp. 88–91. For a contemporary book on the properties of water, cf. Bélidor, *Architecture hydraulique* (see note 11).

22 Perronet shows a knife gate with the function of closing a reservoir in the context of a canal project. Cf. Perronet, *Description*, vol. 2 (1783) (see note 15), plate XLVIII.

23 Voltaire, "Siècle de Louis XIV," *Œuvres complètes de Voltaire*, vol. 4 (Paris: Furne, 1836), pp. 63–309, here p. 222.

f.7 Maison des Directeurs de La Loue; design by Claude-Nicolas Ledoux, published 1804.

Maison des Directeurs de la Loüe.

Vue Perspective

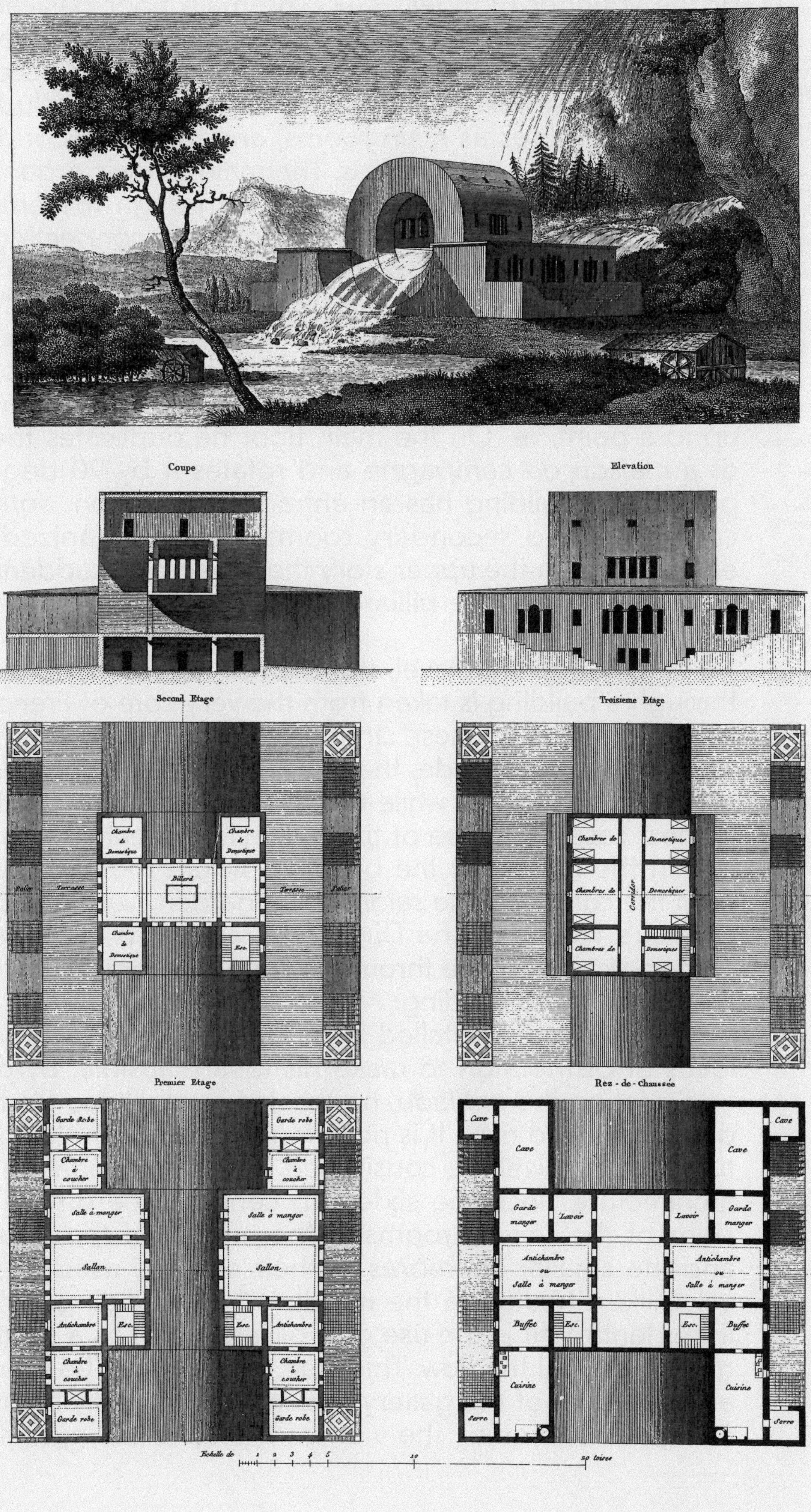

by contemporary authors like Charles Etienne Briseux (1680–1754) or the younger Blondel. 24/f.8 The main floor basically consists of an entrance zone—vestibule and stairs—and several *appartements*. The latter are sequences of rooms to accommodate the owner, his wife, and potential visitors, each including *salons* and/or *chambres* as main rooms, and several secondary rooms like *cabinets* and *garderobes*. The main story is organized along two orthogonal axes: the main axis through the entrance and the *enfilade* along the side of the garden, connecting the main rooms through aligned doorways. The upper stories consist of a number of small *appartements* and *chambres* for visitors or servants. This space is a convenient location for a billiard room, from which the owner and his guests can enjoy the view over the landscape. Ledoux followed this conventional model up to a point. 25 On the main floor he duplicates the structure of a *maison de campagne* and rotates it by 90 degrees. Each part of the building has an entrance axis, *salon*, *antichambres*, *chambres*, and secondary rooms, neatly organized along an *enfilade*. But in the upper story the orientation suddenly changes as the view from the billiard room coincides with the direction of the water flow.

On a general level, the idea of transforming a landscape through a building is taken from the very core of French residential architecture. As these structures communicate between a city-side and a garden-side, they have the task of transforming one sphere into the other while the visitor passes through the interior. In most cases, the area of the highest permeability was the main axis in the middle of the building, where the vestibule opened onto the court and the *salon* to the garden. 26/f.8 In a similar way, Ledoux's House of the Directors transforms wilderness into a garden-like landscape through a zone of selective permeability in the center of the building.

On a more detailed level Ledoux uses the structures and ideas of distribution to make his idea plausible. By duplicating and rotating the *enfilade*, he creates a solid base for taming the dangerous wild river. It is now flanked and controlled by a structure that may well be considered the very foundation of French architecture since the sixteenth century. The different dimensions of the society rooms in the middle of the *appartements* and the smaller *chambres* at their end are used to create the valve-like situation in the middle of the building. Yet the most important motif is the use of a billiard room as a symbolic knife gate to control the flow. This type of space can be considered as a descendent of the gallery, which had originally been intended as a means to enjoy the view of the surroundings while taking

24 Jacques-François Blondel, *De la distribution des maisons de plaisance* (Paris: Charles-Antoine Jombert, 1737–38); Charles Etienne Briseux, *L'Art de bâtir des maisons de campagne* (Paris: Prault, 1743). For the building type and its choice of rooms, see Krause, *Maison de plaisance* (see note 4); for a recent bibliography, cf. Julian Jachmann and Christian Busold, "Typus als Sequenz: Die Maison de campagne bei Charles-Etienne Briseux," *in situ*, 4, no. 1 (2012), pp. 93–102.

25 For other instances in which Ledoux experiments with *appartements*, see Ledoux, *L'Architecture* (see note 3), plates 21, 29. Nicolas Courtin, "L'Ameublement des galeries dans les hôtels parisiens du XVIIe siècle," *Bulletin monumental*, 166, no. 1 (2008), pp. 63–70.

26 Krause, *Maison de plaisance* (see note 4), p. 33.

a stroll during bad weather. In both cases the line of sight from the interior to the outside was crucial, implying pleasure as much as control.

The players in Ledoux's House of the Directors enjoy a view in four directions, yet the axis of the river takes precedent. 27 In addition, two motifs of the topic of mastery over nature are juxtaposed. In Early Modern times, physical activities such as dancing, fencing, or hunting were considered crucial for the education of princes, as they tied the martial roots of the nobility to more recent concepts of body control and politeness important within the context of the court. 28 This idea is employed again in Ledoux's Maison des Directeurs, with the difference that the motif of dominating and cultivating nature—be it the nature of one's own body while playing a game requiring precise movements, or the nature of a torrent—is no longer part of the aristocratic habitus but the self-perception of technocratic professionals. Thus it becomes understandable why Ledoux places a room of seemingly recreational function at this key position. As with his bridge design, Ledoux tries to analyze technical inventions on a structural level, and to include them in the realm of architecture by means of analogy and abstraction. For Ledoux, the central role of water in Chaux, apparent in the Maison des Directeurs as well as technical structures, becomes a stimulus to reflection on the relationship of architecture and machines. Both disciplines have to deal with the qualities of a dynamic and changeable medium: it becomes a *tertium comparationis*.

27 This wish to open up the building towards the stream even leads to a blatant mistake in the pictorial representation of the House of the Directors: the floor plan of the main floor includes three openings of the rooms in the middle axis towards the stream, while the section shows none, thus following the technical possibilities of the time rather than Ledoux's utopian ideas.

28 It was the sociologist Norbert Elias who placed this control of body and emotions (*Affektkontrolle*) at a central place in the history of the nobility and the "process of civilization"; even phenomena like the baroque garden or the famous Machine de Marly can be connected to this concept. Cf. Norbert Elias, *The Court Society*. Trans. Edmund Jephcott (New York: Pantheon Books, 1983 [1969]), pp. 239–46, 259; Elias, *The Civilizing Process: Sociogenetic and Psychogenetic Investigations*. Trans. Edmund Jephcott (Malden, MA: Blackwell Publishing, 2003 [1976]), pp. 60–1. For the position of his theories in recent historical models, cf. Ronald G. Asch, *Europäischer Adel in der frühen Neuzeit: Eine Einführung* (Cologne: Böhlau, 2008), pp. 143–8, 218.

Architecture and machines in the eighteenth century

Even if Ledoux's efforts to address formal problems emerging within the rift between architectural and technical design are undoubtedly as singular as they are curious in outcome, they point to an underlying issue of more universal importance. A central aspect of this topic is a conspicuous convergence of terminology within the fields of engineering and architecture, especially in the central category of *disposition*. In the famous *Encyclopédie* (1751–1772) of Denis Diderot and Jean-Baptiste le Rond d'Alembert several types of machines are mentioned, but the most significant definition is to be found in the article on hydraulic machines by d'Alembert himself. He starts by specifying the general tasks of machines—"à augmenter & à regler les forces mouvantes" 29 —a definition that could directly be applied to Ledoux's faucet-like house. More importantly, a few paragraphs later he emphasizes the Greek etymology of the word "machine," meaning "invention" and "art" rather than "a material object of any kind." This abstract notion of a machine as an ingenious invention

29 Jean-Baptiste le Rond d'Alembert, "Machine," in d'Alembert and Denis Diderot (eds.), *Encyclopédie ou dictionnaire raisonné des sciences, des arts et des métiers*, vol. 9 (Paris: Briasson, 1765), pp. 794–800.

allows the utilization of this word in various heterogeneous fields, such as engineering, painting, and theater.

Returning to the engineer Bélidor, we find the following definition in his *Dictionnaire portatif de l'ingénieur et de l'artilleur* from 1768: "MACHINE. On donne ce nom en général à l'assemblage de plusieurs pieces jointes ensemble, & tellement disposées qu'elles peuveut servir à augmenter ou à régler les forces mouvantes selon les différens usages auxquels on les applique, soit dans la guerre, dans l'architecture, ou dans les autres arts. Il y a des machines simples, & d'autres composées, dont la construction & les usages peuvent se varier presque à l'infini." [30] Again, the regulation of forces, and the need for infinite variation and thus invention are central ideas. But more importantly, they are specified in a way that connects engineering and architecture. [31] The key terms are *disposition* (*disposer*) and *distribution* – both very close in meaning and often used as synonyms. [32] Within architecture these phrases serve to address the spatial structure of a building or ensemble, i.e. the location and position of the buildings and their parts, such as the layout of the ground plan and the structure of the interior space. During the eighteenth century the perfection of these aspects was considered to be a genuine French achievement. Consequently, the terms are ubiquitous in French publications on architecture of this time. [33] Briseux and Blondel call *distribution* a most important art, the latter also a "science inépuisable." [34] Art and science have to meet in the spatial concept, as rooms have to be not only beautiful but also socially appropriate and commodious. [35] Ledoux follows this established line of thought and observes – not unlike Blondel – even a progress in the "science de la distribution." [36]

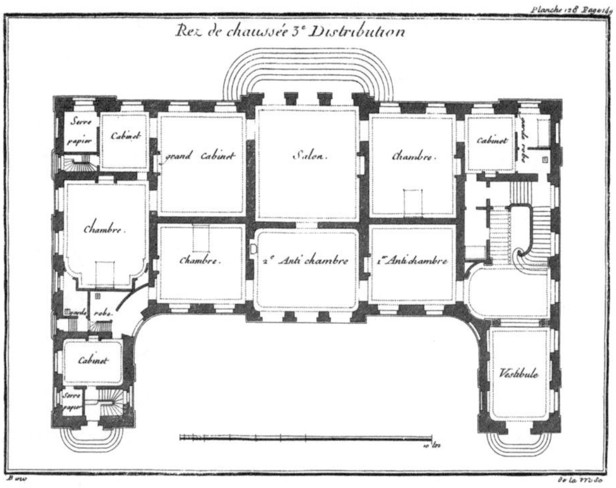

30 Bernard Forest de Bélidor and Charles-Antoine Jombert, *Dictionnaire portatif de l'ingénieur et de l'artilleur* (Paris: Charles-Antoine Jombert, 1768), p. 410.

f.8 In this maison de campagne the middle axis is made permeable, even though the entrance to the building is located in the right-hand wing; design by Charles-Etienne Briseux, published 1743.

31 Another noteworthy correspondence is the term "pièce," which can refer to a single room in architecture as well as to a machine part.

32 There are two antique sources of these terms: one is Vitruvius, the other the theory of rhetoric. Cf. Bettina Köhler, *Die Stadt Paris und das Wohnhaus: Zum "Bâtiment Particulier" in der französischen Architekturtheorie von 1600–1750* (Alfter: VDG, 1994), pp. 207–37; Peter-Eckhard Knabe, *Schlüsselbegriffe des kunsttheoretischen Denkens in Frankreich: Von der Spätklassik bis zum Ende der Aufklärung* (Düsseldorf: L. Schwann, 1972), pp. 123–8. *Distribution* and *disposition* are usually mentioned in encyclopedias and dictionaries, where the huge scope of their application becomes apparent, cf. d'Alembert and Diderot, *Encyclopédie*, vol. 4 (1754) (see note 29), pp. 1062–6; Académie française (ed.), *Dictionnaire de l'Academie française*, vol. 1 (Paris: Smits et Co., 1798), pp. 430, 433 (the definitions in the earlier editions of this work are very similar). Bélidor and Jombert use both words in regard to architecture and engineering. Cf. Bélidor and Jombert, *Dictionnaire* (see note 30), p. 227.

33 Köhler, *Stadt Paris* (see note 32), p. 207.

34 Briseux, *L'Art de bâtir*, vol. 1 (see note 24), p. 27; Jacques-François Blondel, *Cours d'architecture ou traité de la décoration, distribution & construction des bâtiments*, vol. 4 (Paris: Desaint, 1773), p. 189.

35 "La Commodité & la Beauté consistent en la disposition des Pièces," Briseux, *L'Art de bâtir*, vol. 1 (see note 24), p. 18. Cf. Köhler, *Stadt Paris* (see note 32), p. 179.

36 Ledoux, *L'Architecture* (see note 3), pp. 11–12.

The identical terminology made it clear in the eighteenth century that if architecture is mainly seen as an ingenious disposition of elements in space, its design process has to be very similar to that of machines. With this approach the main elements of Vitruvian architecture, such as column orders, decorum, and the triad of *firmitas, utilitas, venustas* are upheld, yet they are subjected to new and dynamic design processes. And vice versa, the organization of machines is sometimes measured by aesthetic standards. In the *Encyclopédie* we find the following remark concerning a machine for the processing of silk: "Le corps de cette machine est simple; c'est aussi sa simplicité qui en fait la beauté." 37

37 D'Alembert, "Machine" (see note 29), p. 798. The central role of "disposition" in the concept of Ledoux is emphasized in Perriault, "Le concept de machine" (see note 1), pp. 115–16.

It could prove an interesting line of inquiry to ask if these terminological convergences explain the sometimes machine-like quality of spatial distribution. **f.9** Apart from the plethora of designs for regular *maisons de campagne* in his publication of 1743, the aforementioned Briseux suggested some unique solutions for difficult sites. One example shows the need of an oblique angle in the middle axis, and the author suggests employing the ordinary layout of two parallel series of rooms but

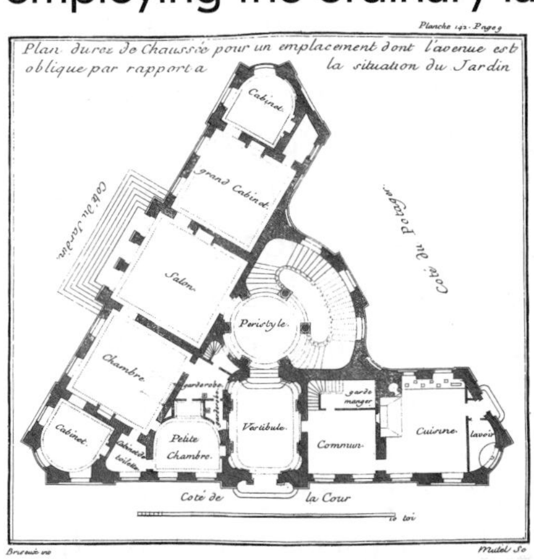

f.9 Triangular maison de champagne; design by Charles-Etienne Briseux, published 1743.

separating them to form a shape like the letter V, with the staircase as the connecting element. The challenging task of bringing the orthogonal structure of rooms and the irregular angle into accord sheds light on the highly flexible and elegant methods used to intertwine spaces of different sizes and layouts, which were established in eighteenth-century domestic architecture. Instead of ideal proportions as defined by Pythagoras and Vitruvius, or a rigid typology, the structure of the building follows the functional concerns of *commodité.* 38 This category must not be confused with modern concepts of functionality, as it describes material functions as well as representative features. The different routes and access points of owners, visitors, and servants within an *appartement* required interwoven sequences of rooms, with secondary rooms like *garderobes* fitted in the remaining spaces. The relationship of the parts of a whole is now to be defined through their importance and functional interdependence. This relationship was necessarily a highly flexible one as rooms in a *maison* or *château* had functions that varied over time, depending on the changing seasons, events, and visitors. This applies in particular to the smaller *maisons de plaisance*, which were not infrequently sold. In order to

38 Köhler, *Stadt Paris* (see note 32).

accommodate not only the different functions of each room but also the changing sequences of ceremonial spaces, a dynamic understanding of the design of floor plans seems to be a commendable strategy, not unlike the approach of machine design, where the alteration of any part influences the whole while the structure in its entirety has to be able to work under different conditions and with potentially varying objectives.

If we ask in this context for possible relationships between architecture and machines, I would like to suggest a different approach than the reduction of architectural thinking to concepts of optimization and the economy of means as employed by Liane Lefaivre and Alexander Tzonis in the 1980s. [39] At least in the eighteenth century, other parallels seem to be far more conspicuous. The central role of *disposition* in the design both of machines and architecture points to an emphasis on a process of invention that is understood as a dynamic one, i.e. as an interrelation of potentially changing parts. Within the field of architecture, this approach manifested itself in the predominant category of *commodité*, which is the complex result of a multitude of material, social, psychological, and semantic constraints, requirements, and ideals—architectonically realized in diverse and highly adaptable floor plans consisting of sequences of intertwined spaces. This rational yet not transparent system is not only employed but also staged. While Ledoux converts in his Maison des Directeurs the middle axis of a typical residential building to transform the torrent and landscape instead of the usual views and functions, Briseux was successful in bringing the main axes and spatial sequences into accord with the oblique angle of the site. And in a completely rational way the building serves to connect and organize three outside areas: court, garden, and kitchen garden. To emphasize the hinge-like structure of the floor plan, Briseux employs the round *peristyle* in the center of the building. He thus not only makes his idea of rotating one of the two room sequences visible, but also adds a dynamic characteristic, underlining the diverse possibilities and changing directions of design processes.

39 Liane Lefaivre and Alexander Tzonis, "Die Maschine im architektonischen Denken / The Machine in Architectural Thinking," *Daidalos*, 18 (1985), pp. 16–26; Lefaivre and Tzonis, "The Mechanization of Architecture and the Birth of Functionalism," *Via*, 7 (1984), pp. 121–43.

The Museum in Motion: Two Projects by Made in

Tobias Erb

Today, the art museum is trapped between its desire for a distinctive visual profile for which the architect takes authorship, and the obligation to respect a whole range of programmatic specifications drawn up by anonymous specialists. Correspondingly, the task of the architect during the design process remains categorically distinct from any challenging curatorial, institutional, or technical demands. In its recent design competition entries for two prominent museums in Switzerland, the Geneva-based architecture office Made in called into question this state of affairs and ultimately renegotiated the competition brief precisely in this regard. Its proposals for the new Musée Cantonal des Beaux-Arts Lausanne and an extension to the Kunstmuseum Basel evince a new angle on the art museum, treating it not from the traditional perspective, as a private collection or scholarly cabinet, but rather as a locus of collective experience. They understand the art museum as an apparatus that actively generates knowledge and explore the alternative ways in which art works might be staged.

In Lausanne, this culminated in a novel layout for the new Musée Cantonal des Beaux-Arts. The exhibition space comprises four cylindrical towers set at the corners of a public park. Rather than floors, each tower is equipped with a lifting platform that vertically transports visitors from the ground-level entrance through the exhibitions, whereby the permanent collection is displayed on the walls and temporary exhibitions on the lifting platform itself. Once at the top, visitors take exterior elevators straight back down to the park before proceeding to the next tower to repeat the process. This alternative to the classic horizontal perambulation is not only a literal interpretation of the program specifications, such as to provide the greatest ratio of wall space to floor space, but also fundamentally alters the exhibition and viewing scenarios. On the one hand the

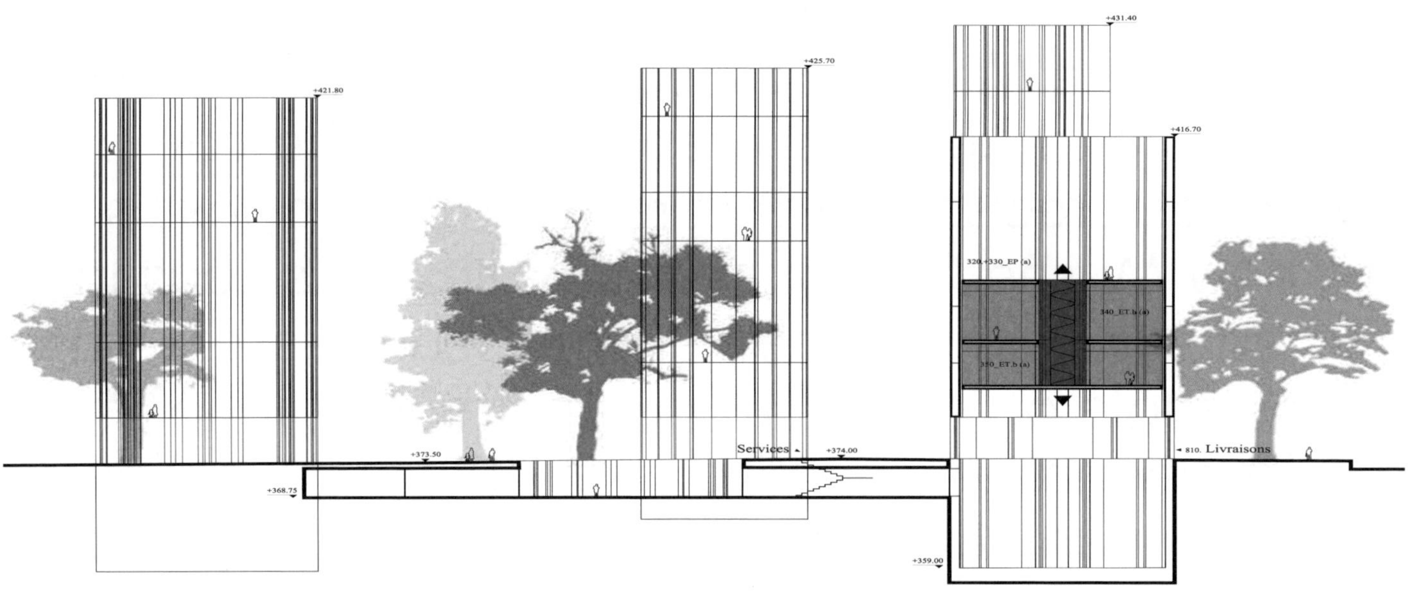

movement of the lifting platform enables the permanent collection to be viewed in individual scenarios, and on the other sets it in a shifting relationship to the temporary exhibitions.

A similar principle underpins Made in's proposal for an extension to the Kunstmuseum Basel, where the architects understood their remit as being not only to enlarge the museum, but also to expand its potential. The existing building, with its traditional enfilade for the permanent collection, was complemented by the addition of a new one for temporary exhibitions. This consists of modular wall and floor elements suspended from a large roof construction, and can thus be flexibly configured as rooms, galleries, or cabinets, enabling new forms of staging to be explored. Visitors' movement between these two self-contained buildings assures the unity of the permanent collection and the temporary exhibitions, as required by the program. The primary access point is still the entrance to the original building, from where it is possible to undertake a tour of the various sections, the tour being articulated by a series of escalators, lifts, or revolving doors. These elements are so arranged as to allow the works of art and the exhibitions to be placed in different relationships to one another, thus opening up shifting scenarios.

Although the architectural dispositions of the two projects differ, they nevertheless have a core factor in common, namely the involvement of mechanical elements in the form of lifting platforms, escalators, elevators, or revolving doors. These serve to explicate the idea of the museum as an apparatus: not merely by directly facilitating constantly new encounters between the viewer and the object, but also superordinately, inasmuch as they lend the art museum a new dimension of temporality. The primary emphasis of the design shifts therefore to performance and potentialities—to a museum that is judged not so much by its appearance as by the availability and range of its scenarios. Made in's contributions to the competitions for the new Musée Cantonal des Beaux-Arts Lausanne and the extension to the Kunstmuseum Basel negotiate not only a new architectural paradigm for the art museum per se, but also a new perspective for architects. They see the definition of the programmatic specifications as part of the design process, thus tapping into the architectural potential embodied in the competition brief while simultaneously transcending the role commonly prescribed for architecture in it. The specialists may draw up precise specifications, but it is the task of the architect to invest them with meaning.

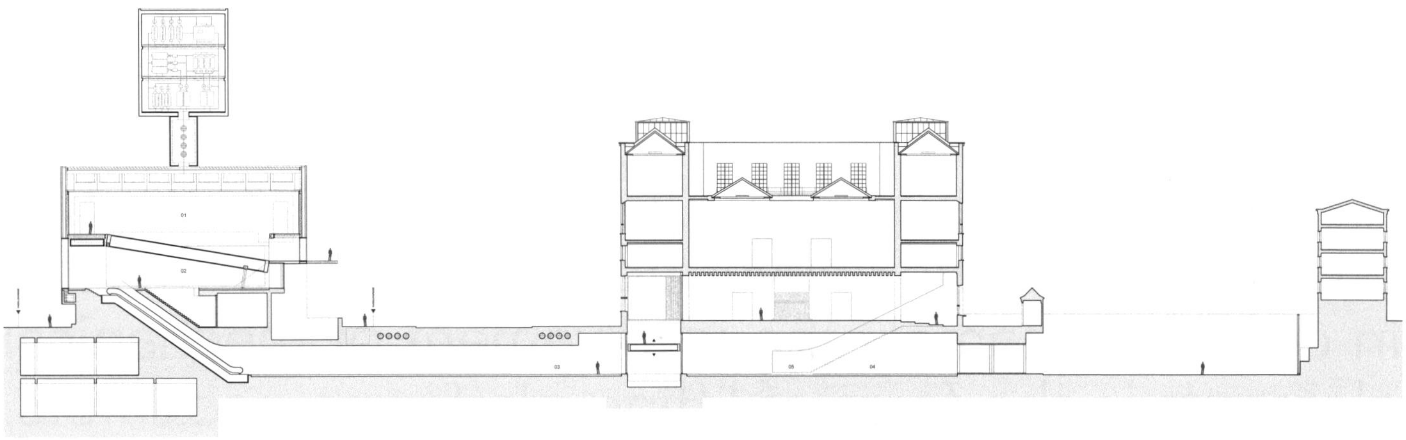

f.1 Made in, Geneva, new Musée Cantonal des Beaux-Arts Lausanne, competition 2004, section.
f.2 Made in, Geneva, extension Kunstmuseum Basel, competition 2009, section.

Viollet-le-Duc's Organic Machine

Martin Bressani

In his last didactic children's novella, *Histoire d'un dessinateur: Comment on apprend à dessiner,* published posthumously in December 1879, the celebrated French neo-Gothic architect and theoretician Eugène Emmanuel Viollet-le-Duc illustrated a mechanized artificial limb, showing how the workings of muscles and tendons in animals can be used as model for mechanical devices. f.1 It served as conclusion to two successive anatomy lessons featuring the two main protagonists of the short novella—the industrialist Majorin and his young protégé petit Jean. Through the study of anatomy, explains Majorin to his pupil, one can gain a new appreciation of machines in factories, "for man, in the arts of mechanization, seldom does more than apply these anatomical elements." He adds that "one could write a whole treatise on mechanics, taking as the sole subject the curvature of the bones," and goes on to lament that man will never be able to achieve the ultimate goal of producing a creature with the same level of autonomy as a living being. But by using organic models to shape mechanisms, Viollet-le-Duc hints at a possible continuity between machines and humans. As he describes it: "the organs of [man's] best machines are generally made in conformity to the principles by which his body moves." 1

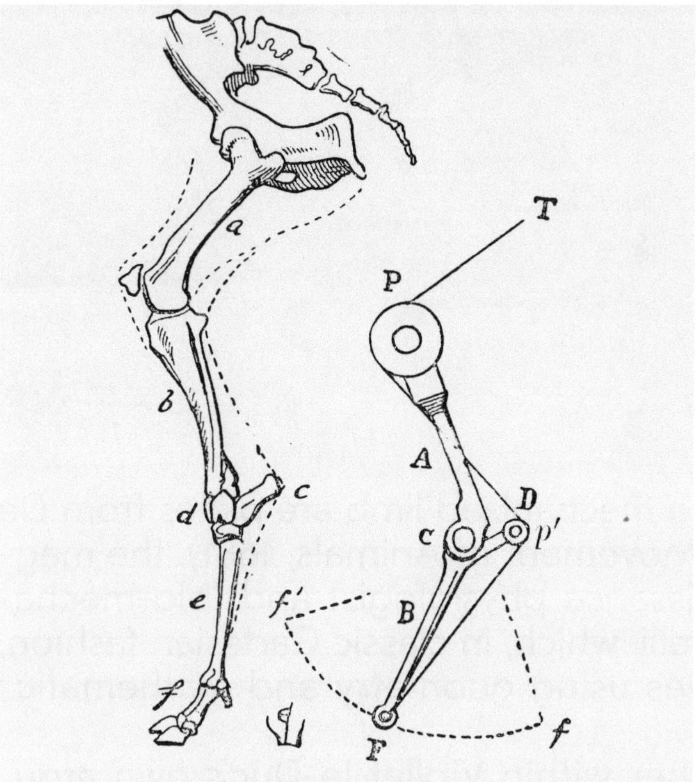

Fig. 60. — Application du jeu des muscles et tendons à la mécanique.

A reversal of the traditional (Cartesian) relationship between Body and Machine, Viollet-le-Duc's striking drawing is, to my knowledge, unique in the annals of nineteenth-century technical illustrations. We are of course familiar with the mechanist explanation of organic phenomena from the long-standing Cartesian tradition, but the reverse idea—that machines should be modeled upon organisms—is highly unusual before the twentieth century. It is worth pondering such early manifestation of the coupling of machine and organism in the work of one of the

Martin Bressani is Professor and Director of McGill University's School of Architecture.

Note: This essay is the last of a trilogy of texts I have written on the *Histoire d'un dessinateur,* which partly overlap but complement each other. The two others are 1, the concluding chapter of my *Architecture and the Historical Imagination: Eugène-Emmanuel Viollet-le-Duc 1814–1879* (Farnham [Surrey, UK]: Ashgate Publishing Limited, 2014), pp. 499–519; and 2, the article "Prosthetic Fantasies of the First Machine Age: Viollet-le-Duc's Iron Architecture," *AA Files,* 68 (2014), pp. 43–9.

f.1 Eugène Emmanuel Viollet-le-Duc. The working of muscles and tendon applied to a mechanical device. Woodcut engraving.

1 [Eugène Emmanuel] Viollet-le-Duc, *Histoire d'un dessinateur: Comment on apprend à dessiner* (Paris: J. Hetzel, [1879]), pp. 131–5. English translation by Virginia Champlin, *Learning to Draw; or, The Story of a Young Designer* (New York, NY: Putman, 1881). All translations from *Histoire d'un dessinateur* are based on this edition, but I emended them whenever I felt it was necessary in order to render more accurately Viollet-le-Duc's original meaning. All further translations are my own.

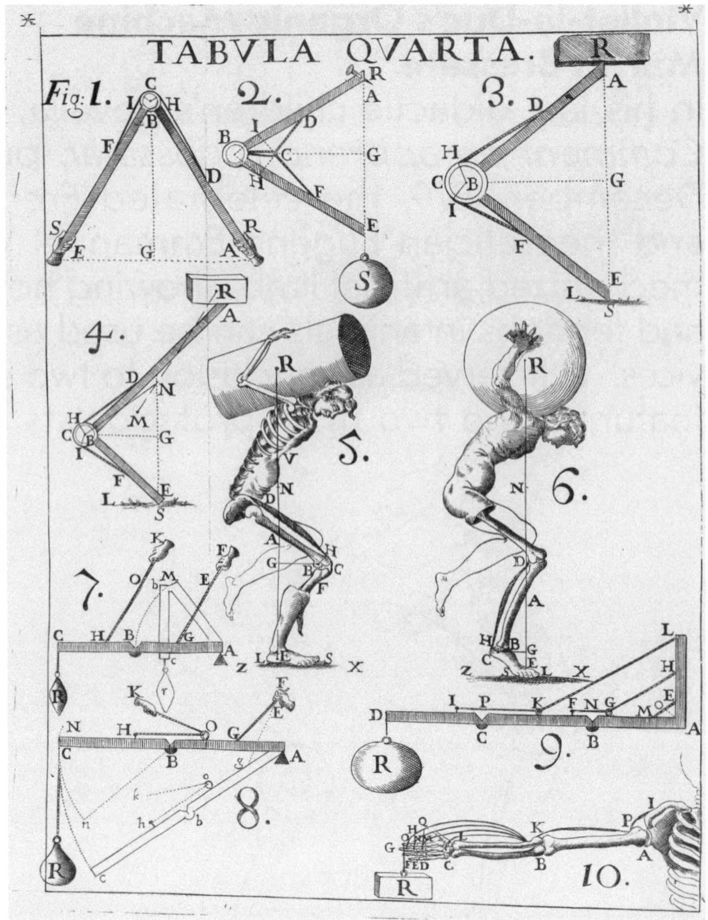

f.2 Plate IV of Giovanni Alfonso Borelli's *De Motu Animalium*, 1680.

key theoreticians of modernism. Is Viollet-le-Duc anticipating the destabilization of the boundary between the human and the non-human that has been often associated with post-Second World War developments and the concept of cybernetic organisms? [2] Or is his mechanized limb simply a confusion of terms, a late avatar of the traditional Cartesian theory of the animal-machine? After all, in the same novella, Viollet-le-Duc does describe the human body as a machine mechanism. [3] And the closest precedent to his image of a mechanized limb are plates from *De Motu Animalium* (On the Movement of Animals, 1681), the magnum opus of Italian Renaissance physiologist and "bio-mechanist" Giovanni Alfonso Borelli, which, in classic Cartesian fashion, relates animals to machines using geometry and mathematics to prove his theories. [4/f.2]

Before delving further within Viollet-le-Duc's own argument to shed light on these questions, it is useful to first establish if there were other nineteenth-century instances of this reversal of the traditional Cartesian relation between mechanism and organism. To be sure, poetic fantasies about machines, comparing locomotives to dragons and other beastly forms, were common in nineteenth-century literature. [5] And British novelist Samuel Butler, in his famous dystopian novel *Erewhon; or, Over the Range* (1872), described man "as a machinate mammal." [6] But these literary takes on the world of mechanization, while significant in their own right, are of limited use if we seek to understand the conceptual turn that led to the reversal of the Cartesian analogy of organism/machine.

In a well-known lecture titled "Machine and Organism" dating from 1947, French philosopher and physician Georges Canguilhem provides us with useful insights for tracking conceptual sources. He explains that the reversal first originated in inquiries attentive to the relationship between the production

2 See, for instance, Andy Pickering, "Cyborg History and the World War II Regime," *Perspectives on Science*, 3, no. 1 (1995), pp. 1–48.

3 Viollet-le-Duc, *Histoire d'un dessinateur* (see note 1), p 128.

4 Viollet-le-Duc had a precious edition of Borelli's famous book in his library: *De Motu Animalium*, Editio novissima auctior et emendatior (Hagae Comitum: Apud petrum Gosse, 1743). See item 1134 in *Catalogue des livres composant la bibliothèque de feu M. E. Viollet-le-Duc* (Paris: Labitte, 1880), p. 148.

5 See my extended discussion of these poetic fantasies in *Architecture and the Historical Imagination* (see preliminary note), pp. 415ff.

6 Samuel Butler, *Erewhon; or, Over the Range* (London: Trübner, 1872), p. 221.

of primitive tools and organic activity itself, in other words, in reflections on the historical sources of technological phenomena. It is through such an "ethnological" outlook, claims Canguilhem, that tools and, by extension, machines came to be considered as prolongations of the organs of the human species. He identifies Ernst Kapp's theory of "organ projection," developed in the latter's *Grundlinien einer Philosophie der Technik* (1877), as the initiator of such organic understanding of technology. Within the French context, he points to philosopher Alfred Espinas's *Les origines de la technologie* (1897) as the first to have followed Kapp's lead. [7]

If Kapp and Espinas were indeed pioneers in elaborating full-fledged theories of tools or machines as organic projections, they were not the only ones to have given expression to the concept in the nineteenth century, as the *Histoire d'un dessinateur* already bears witness. Using Viollet-le-Duc's work and its own complex network of sources as a thread, this article will serve to further flesh out Canguilhem's classic essay of 1947 by expanding its insights in several directions—first by identifying ethnographic precedents, as he suggested, but also other sources in Romantic vitalist philosophies, neo-Catholic doctrine, and anatomical studies.

The earliest nineteenth-century expression of a biological take on human technique—some version of the idea had already been articulated in the seventeenth century [8] —is probably the short sentence (quoted by Kapp) at the end of the second volume of Alexander von Humboldt's illustrious masterpiece *Kosmos: Entwurf einer physischen Weltbeschreibung,* published in 1847: "The creation of new organs or instruments of observation," writes Humboldt, "augments the intellectual, and often the physical powers of man." [9] Humboldt's immensely popular five-volume panorama on science and the universe sought to establish, in classic Romantic fashion, a unifying perspective on the studies of science, nature, and mankind. Its impact was immense, including in France, where it was translated almost immediately. It may have influenced the work of pioneer prehistoric archaeologist Jacques Boucher de Perthes, the first in France to clearly express the idea that tools are extensions of human organs. In *Des outils de pierre,* a pamphlet published in 1865 on Stone-Age tools, he asserts that primitive tools were "a consequence of [man's] physical constitution, a sort of addition to ourselves or a supplementary organ [*membres supplémentaires*]," [10] a claim which he illustrates with plates showing how primitive flints fitted exactly the contour of the hand and thus prolonged and extended the arm's organic movement of percussion. The remark

7 Georges Canguilhem, "Machine and Organism," in Canguilhem, *Knowledge of Life.* Trans. Stefanos Geroulanos and Daniela Ginsburg (New York: Fordham University Press, 2008), pp. 75–97; here p. 93.

8 The idea that scientific instruments (not tools per se) are extensions of natural human organs may have been a current assumption within the Royal Society in seventeenth-century England, at least judging from the following passage in Robert Hooke's *Micrographia* of 1665: "The next care to be taken, in respect of the Senses, is a supplying of their infirmities with *Instruments*, and, as it were, the adding of *artificial Organs* to the natural;" Robert Hooke, "Preface," in R[obert] Hooke, *Micrographia; or, Some Physiological Descriptions of Minute Bodies Made by Magnifying Glasses with Observations and Inquiries Thereupon* (London: Jo. Martyn and Ja. Allestry, 1665), n.p.

9 Alexander von Humboldt, *Cosmos: Sketch of a Physical Description of the Universe*, vol. 2 (London: Longman, Brown, Green & Longmans, 1848), p. 358.

10 Jacques Boucher de Perthes, *Des outils de pierre* (Paris: Jung-Treuttel, 1865), p. 6.

was far from casual, as Boucher de Perthes had earlier constructed a vast "organology" of the world, developing the idea that human and animal organs are projections of the will, and therefore subject to improvement and metamorphosis. He expounded these ideas in his first substantial work, *De la creation: Essai sur l'origine et la progression des êtres* (1838–1841) — a curious five-volume philosophical panorama of the universe which analyses organ formation within the universal metamorphosis of matter towards the creation of organisms with ever increasing perfection. But beyond the brief passage quoted above in *Des outils de pierre*, he never pursued the idea of man's industrial powers in terms of an organology.

The idea is more fully developed in an apology of industry written in 1884 by a disciple of Auguste Comte, the polymath and historian Louis Bourdeau, in which machines are described as "the equivalent of a new natural kingdom" and "counterfeits of animated beings." [11] This new mechanical order, writes Bourdeau, allows man to adapt his "bodies to functions that nature had not planned." It "completes our still unfinished organism. [Machines] are like auxiliary organs that we can wear or take off, exchange and modify at will." [12] In short, claims Bourdeau, "man reproduces and summarizes with his technical artifices all the scattered perfections of the animal kingdom," an observation that could well have served as caption to Viollet-le-Duc's mechanized limb modeled on the hind leg of a cervid. Though indebted to Comte's sociology, Bourdeau's work is not unrelated to the romanticism of Humboldt and Boucher de Perthes insofar as, like them, he sought to establish a general philosophical synthesis of nature in which to place man. His most ambitious philosophical work, *Le problème de la vie: Essai de sociologie générale* (1901) was an attempt at a monistic science that sought to identify a general principle of life active in both the natural and human world.

This brief and no doubt very partial account of early expressions of the organic understanding of machines, especially in France, clarifies the structure of thought and attitudes from which the reversal of the traditional Cartesian relation between mechanism and organism could be effected. Confirming Canguilhem's insight, it was by considering the unconscious creative impulse behind the generation of machines, or their "irrational origin" to use Canguilhem's own expression, that technique could come to be understood as a biological phenomenon. [13] Once conceived as a natural behavior of the living, or of life in general, machines naturally aligned themselves, so to speak, with the functioning of the human body. They became its extension, or its "organ projection," to use Kapp's stunning formulation.

11 Louis Bourdeau, *Les forces de l'industrie: Progrès de la science humaine* (Paris: Félix Alcan, 1884), pp. 233–4.

12 Ibid., pp. 45–6.

13 Canguilhem, "Machine and Organism" (see note 7), p. 95.

There is some basis to suppose that Viollet-le-Duc was acquainted with the work of Humboldt and Boucher de Perthes, if not Bourdeau. 14 But whether he knew these authors or not, we can trace the same set of attitudes in his own life-long inquiry into the nature of creativity at work simultaneously in both the natural and human world. His theory of *style*, which is his term to denote such creative impulse, is couched in a celebration of nature's generative principle, demonstrating how authentic (as opposed to degenerate) human creation must be understood by way of analogies with the formative powers of nature. In the article "Style" of his great *Dictionnaire raisonné de l'architecture française du XIe au XVIe siècle* he refers, at some point, to the machine as the paradigm of an object endowed with style—"the machine is the exact expression of the function it fulfills; us, artists, need not go any further" 15 —but the examples he uses are the wings of a bird and the curve of the body of a fish. It is not an abstract engineering logic that matters to him, but a natural, organic process of formation. To be legitimate, human creations must be the product of a true creative act, a pure expression of the will. 16 Despite being so often categorized as a rationalist, Viollet-le-Duc, who was above all an archaeologist given to general reflections on man's historicity, believed that authentic human technique stemmed from a vital originality irreducible to pure rationalization. He had no patience with the new mathematization of technical knowledge by modern engineers, while he took great care, as we will see, to study the gesture and body postures of craftsmen, soldiers, and other men of action, particularly from the Middle Ages, his privileged domain of historical inquiry.

14 Viollet-le-Duc greatly admired Alexander von Humboldt, as he himself states in several publications (see [Eugène Emmanuel] Viollet-le-Duc, "Antiquités américaines," in Désiré Charnay, *Cités et ruines américaines* (Paris: Gide et Morel, 1863), pp. 3–104; here pp. 3–4, and E[ugène Emmanuel] Viollet-le-Duc, *Mémoires sur la défense de Paris: Septembre 1870–Janvier 1871* (Paris; Morel, 1871), p. 72. It is extremely likely that he was acquainted with the work of archaeologist Boucher de Perthes, who was close to his own father and to Napoleon III. As for the young Louis Bourdeau, he may have met him in the positivist circles in which he moved in the 1870s—though that is highly conjectural.

15 [Eugène Emmanuel] Viollet-le-Duc, *Dictionnaire raisonné de l'architecture française du XIe au XVIe siècle*, vol. 8 (Paris: Morel, 1866), p. 488.

16 See my extended discussion of Viollet-le-Duc's idea of "style" in *Architecture and the Historical Imagination* (see preliminary note), pp. 381–92.

Anatomy as a *science-principe*

What distinguishes Viollet-le-Duc's juxtaposition of organism and machine from the authors just reviewed, however, is that in the *Histoire d'un dessinateur* it assumes the status of a method, a precise and functional exchange between organism and machine. The industrialist, according to Majorin, must study anatomy to create his machines, animal limbs serving as direct inspiration for machine mechanisms. The fact that Viollet-le-Duc chose to illustrate how mechanism can be modeled upon animals is highly significant: the organic was to form part of the imaginary of future industrialists. For Viollet-le-Duc, who enjoyed the work of old Renaissance masters and scientists, such as Leonardo (whom he discusses directly in the *Histoire d'un dessinateur*), 17 and Borelli, whose curious anatomical treatise on the movement of animals he purchased, 18 anatomy had acquired

17 Viollet-le-Duc, *Histoire d'un dessinateur* (see note 1), pp. 205–8.

18 See note 4.

the status of a *science-principe*. It was a primordial resource for those who like to ponder the hidden springs of the universe, an ancient knowledge that seeped through time through the work of kindred spirit.

That ancient knowledge formed of course the backbone of his earlier writings on architecture. In his ten-volume *Dictionnaire raisonné de l'architecture française,* Viollet-le-Duc developed a biological idiom for architecture. In the preface to volume one dating from 1854, Viollet-le-Duc affirms "that the moment has arrived to study the art of the Middle Ages like one studies the development and the life of a living being [*la vie d'un être animé*]." [19] His magnum opus is indeed an anatomy treatise of medieval architecture, with its thousands of illustrations dissecting the "organs" of monasteries, castles, and especially cathedrals. Viollet-le-Duc compares the latter's "interior *organization*" to the human body, "complexifying and developing as it progressively contains a greater number of crucial organs." [20] Viollet-le-Duc renewed with the Vitruvian body metaphor but corrected its meaning in claiming that the translation of the human body into architecture was no longer conceived as the product of an imitative process, as in the classical tradition, but of an instinctive projection of man's bodily organization.

His own sources for the idea that architecture, or art more generally, is a form of incorporation can be traced to the ideological foundations of the French neo-Gothic movement, most notably to the writings and teaching of physician, philosopher, historian, and socialist Christian, Philippe Joseph Benjamin Buchez, whose "school" met next door to Viollet-le-Duc's family house on rue Chabanais in Paris. And Buchez's work sheds an interesting light on the question of an instinctive "body projection." As many other proponents of a return to *l'art chrétien*, he saw human creation as a form of incarnation modeled on that of Jesus. Man's highest creative act proceeded through an overwhelming sense of a unified bodily consciousness, obliterating memory and habit while disclosing a new historical truth. [21] Viollet-le-Duc was exposed to these ideas very early through his participation in the Institut historique de Paris in the 1830s, a learned society dominated, from the start, by Buchez's school. [22] But he probably assimilated Buchez's teaching more thoroughly through his friendship with the anatomist and medical doctor Jean-Marc Bourgery, himself probably a disciple of Buchez (whom he quotes repeatedly), and author of the monumental *Traité complet de l'anatomie de l'homme* (1831–1854), generally recognized as the most lavish human anatomy treatise produced in the nineteenth century. The heavy tomes were by far

19 [Eugène Emmanuel] Viollet-le-Duc, *Dictionnaire raisonné de l'architecture française du XIe au XVIe siècle*, vol. 1 (Paris: Bance, 1854), p. vi.

20 Ibid., p. 149. Emphasis in original.

21 Philippe Buchez, *Introduction à la science de l'histoire ou science du développement de l'humanité* (Paris: Paulin, 1833), pp. 194–5.

22 For an extended discussion of Viollet-le-Duc's ties to Buchez, see my doctoral dissertation, "Science, histoire et archéologie. Sources et généalogie de la pensée organiciste de Viollet-le-Duc" (PhD, Université de Paris-Sorbonne Paris IV, 1997), pp. 153–68.

the most prominent work on natural sciences in Viollet-le-Duc's own library. An attentive reading of the *Traité complet* draws the reader into the convolutions typical of Romantic biology. Bourgery opens his *Traité* by establishing the model of the ideal human body—a normative model which, as I have argued elsewhere, 23 is not unlike Viollet-le-Duc's "ideal" cathedral in his *Dictionnaire raisonné de l'architecture*. 24 Following in the footsteps of Buchez, the real body is the model, but a body seized at its maximum level of vitality, product of the perfect equilibrium between the material and the spiritual. Not surprisingly, Bourgery's text integrates the notions of instinct and race as determinant agents. "It is through his spiritual self, transmissible through the race, that are produced, thanks to the collective work of generations, all great manifestations of the spirit." 25 And just as we have seen in Buchez (and Viollet-le-Duc), Bourgery would be led to see human creations as a form of exteriorization of the body, in a way which anticipates Ernst Kapp's theory of projection:

"In all applications of the mind within the external world, the organism does but copy itself, transposing itself, as it were, and imposing itself upon nature. ... [M]an does not precisely invent anything, since he imagines only what he senses within himself, applying to the outside what he is, or what he has been made inside." 26

23 See Bressani, *Architecture and the Historical Imagination* (see preliminary note), pp. 277–88.

24 Jean-Marc Bourgery, *Traité complet de l'anatomie de l'homme*, vol. 1 (Paris: Delaunay, 1832), p. 3.

25 Ibid., vol. 3 (1844), p. 18.

26 Ibid., p. 24.

The soldier-weapon

Despite obvious continuities, we need to be aware that there is a significant distance between Viollet-le-Duc's reflections on architecture from the 1850s and 1860s and the mechanized limb he illustrated in 1879. He had relatively little interest in mechanization before the late 1860s, if we except his great fascination with medieval war machines, which he describes in great detail over 50 pages in volume five of his *Dictionnaire raisonné de l'architecture* (1861). The shift from architecture to mechanization seen in the *Histoire d'un dessinateur* was not a casual swing of interest, but reflected a changed perspective. The key factor for this transformation was the disastrous Franco-Prussian War (1870–1871), the first military conflict, together with the American Civil War and the Austro-Prussian War, to have clearly displayed aspects of industrial warfare. Viollet-le-Duc was fully conscious of that new reality, as his fascinating *Mémoires sur la défense de Paris: Septembre 1870–Janvier 1871* bears witness. An exposition of his thoughts on what France should have done (and should do in the future) to win the war against Germany, the *Mémoires* lay out the new requirements of modern warfare. It is now a vast "enterprise," described Viollet-le-Duc, which, like industrial production, requires an efficient administrative organization, a knowledge of how to gain and use capital, and control of the largest number of weapons and the most complete array of resources:

"It is not merely the reorganization of the army that must now be envisaged, but the organization towards battle of the entire forces which a country like France has at its disposal, the potential transformation, at any given time, of all the peacetime productive forces into the destructive forces of warfare. ... Industry—and this was proven [in the Franco-Prussian war]—produces as easily war engines as instruments for peace. The special knowledge of engineers, geographers, physicians, chemists, must be used equally for the one and the other." [27]

From there on, the ardent patriot in Viollet-le-Duc would focus on education, and seek to reach out directly to "the children of all classes," as he himself put it to his publisher Jules Hetzel, teaching the young people of France how to think, reason, and muster the energy "to get up in the morning and set to work." [28] The *Histoire d'un dessinateur* is the last and ultimate product of his series of pedagogical publications aimed at such regeneration of the youth. Despite its seemingly benevolent tone, the short novella is entirely couched in this obsessive desire to redress the French nation—an ambition that followed quasi-naturally from his life-long neo-Gothic quest, but now redirected towards industry and France's military future.

In this regard, it is useful to review at this point the novella's basic story line. Petit Jean is an 11-year-old country boy, son of père Loupeau, a hard-working gardener and handyman living on the outskirts of Paris. One day the eccentric industrialist, factory-owner Monsieur Majorin, pays a visit and notices a drawing made by petit Jean of his cat. Detecting an unusual level of aptitude, he offers to take full custody of the boy and make him his pupil. Ripped away from the bosom of his natural family as if entering military training, petit Jean is subsequently drilled in various subjects over several chapters. Especially fundamental to this pedagogical experiment is the development of his ability to draw, through which the reader understands he is learning how to see, and therefore to think and reason. Viollet-le-Duc is explicit about what this discipline of seeing entails: the interrogation of one's surroundings in order to rebuild them mentally from the ground up:

"When you see any thing, a piece of furniture, a tool, or a house, you must ask how it is made, with what, and why, and try to guess it yourself. ... When you see an animal, big or little, a horse or a sheep, or an insect or a bird, you must ask what it

27 Viollet-le-Duc, *Mémoires sur la défense de Paris* (see note 14), pp. 226–7.

28 Passage from a letter from Viollet-le-Duc to his publisher Jules Hetzel, July 20, 1874, Médiathèque de l'architecture et du patrimoine, Paris, "Correspondance et rapports, 1872–75," doc 93.

does in order to walk, to defend itself, to get sustenance, and to fly. When you see a plant, you must ask how it comes up from the earth, how it pushes forth shoots; and watch closely, and see how the leaves, flowers, and fruit are attached." 29

For petit Jean, the act of seeing and drawing thus becomes a relentless anatomical analysis, an act of reconstruction, as if the world has to be pulled back from the brink of self-destruction and made anew. It was not towards architecture or the arts that Majorin's drawing pedagogy was leading petit Jean, but towards industry or, more generally, towards the development of a new capacity for action. Whereas the Second Empire had engendered a fragmented, decentered self, the youth of France of the Third Republic had now to cohere into a new unity, ready to take control of his environment. And with this control, the passive body of the Second Empire had to give way to a new self, transformed into an active and intelligent part of the larger body of a (regenerated) nation. The development of new organs, in this case visual, was necessary to construct this new living apparatus.

When seen in this context, the reach of Viollet-le-Duc's mechanized limb illustrated in *Histoire d'un dessinateur* is extended quite considerably. Petit Jean's comparative anatomy lessons, which end on the topic of mechanization, come immediately after the crucial chapter titled "Petit Jean Begins to See," in which the young apprentice reached a first degree of awareness of the vitality that runs throughout nature. Only then can the anatomy lessons begin, looking at two rather terrifying creatures—a bat and a Pterodactyl—thus setting the theme of nature's great variety with an undertow of monstrosity before he moves to human anatomy. Once the secret of anatomy is disclosed, petit Jean can gain a more conscious understanding of the productive force of nature and thus be able to see how industrialization can fully assume its role as the creator of a second nature. The organs of men's machines, declares Majorin, must now be made in conformity to the principles of his own body. Majorin describes the human hand as a tool, perfected through evolution, and the human skeleton as a mechanism. 30 It is not a return to the old Cartesian paradigm, as the issue is now to link the human body to the machine—a question naturally tied to a first industrial age when the machine progressively took over the human habitus. 31 Viollet-le-Duc, however, does not conceive the factory as a place of repetitive gestures and nascent Taylorization, but as a living, adaptable apparatus. At any given time, he repeats, France's "civilian organization must be able to transform, within twenty-four hours, into a military organization." 32 As a national center of production, industry must be adaptable

29 Viollet-le-Duc, *Histoire d'un dessinateur* (see note 1), p. 21.

30 Ibid., pp. 115, 119, and 128.

31 On the link between architecture and eugenics see Georges Teyssot, "Figuring the Invisible," chapter two of his *A Topology of Everyday Constellations* (Cambridge, MA: MIT Press, 2013), pp. 31–82.

32 Viollet-le-Duc, *Mémoires sur la défense de Paris* (see note 14), p. 231.

A

and vital, a capacity put to the test in times of war. War is indeed the vital regenerator, "the sole agent," writes Viollet-le-Duc, "capable of conserving the moral energy which gives nation its power and cohesion." 33

f.3 Eugène Emmanuel Viollet-le-Duc. Fifteenth-century armor. Woodcut engraving.

33 [Eugène Emmanuel] Viollet-le-Duc, *Dictionnaire raisonné du mobilier français de l'époque carlovingienne à la Renaissance*, vol. 5 (Paris: Morel, 1874), p. 3.

In the same way, he describes how the modern soldier has become "nothing but a force transmitted to a weapon; the more that weapon will be perfected ... the more man will be reduced to the state of a trigger mechanism." Yet an absolute mechanized conception of warfare is despicable to him, an ignobility that he associates with the German army. A superior military training assumes that soldiers have "personalized" their weapon, bringing their own "intelligence, prevision, knowledge" into play. Only then can be ensured the "superiority to that horrifying mechanism which we call a weapon." 34 The mechanism, in other words, must always find its place back into living, adaptable, organic life.

34 Ibid., p. 5.

Such intimate and vital relationship between soldier and weapon is the subject of the entire fifth and sixth volumes of Viollet-le-Duc's remarkable *Dictionnaire raisonné du mobilier de l'époque carlovingienne à la Renaissance* (1874 and 1875),

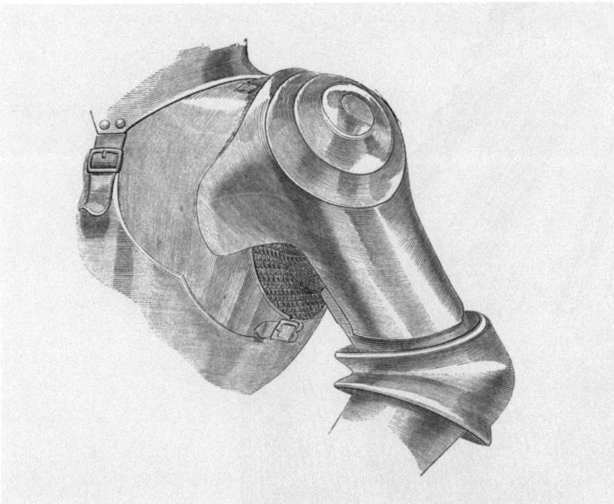

f.4 Eugène Emmanuel. Viollet-le-Duc. Fifteenth-century spaulders. Woodcut engraving.

a study of medieval weaponry published just a few years before *Histoire d'un dessinateur*. The volumes were written with the same goal of national regeneration as his didactic children's novels: "If this last part of our work ... can contribute to impress upon the mind the love of the craft of war," wrote Viollet-le-Duc in the preface to volume five, "if it can demonstrate how, after so many unprecedented disasters, France has been able, thanks to patriotism, to erase many of its errors and to recover, we think we will have fulfilled a small part of the task that each Frenchman must carry today." 35

35 Ibid., p. 9.

As always, within Viollet-le-Duc's historicized outlook, the Middle Ages stand paradoxically as the fantasy reservoir for France's future. Medieval knights—"these men covered in iron" 36 —were models for the "hardened" soldiers of the future. f.3 In fact, Viollet-le-Duc transposed the entire modern militarized state onto the Middle Ages: feudal society, thanks to its perpetual state of warfare, had given birth to new iron creatures, just as nineteenth-century industrial Europe was in the process of engendering a new type of mechanized being. Much of Viollet-le-Duc's

36 Viollet-le-Duc, *Dictionnaire raisonné de l'architecture* (see note 19), vol. 3, p. 164.

wonderful and manically detailed history of medieval armament centers on the adaptation of body to weapon, weapon to body. f.4/f.5 Hundreds of fascinating illustrations depict medieval warriors as organisms that "evolve" into a protective shield, speculating that medieval armorers modeled their work on the shells of animals or insects. 37 His essential point is that armor was forged through a very delicate observation of the disposition of the play of muscles and physical movement. 38 Through comparative illustrations he demonstrates how certain parts are adapted to human anatomy f.6 to produce responsive "metallic military clothing," garments that were all the more astonishing given "that medieval industrialists didn't possess the anatomical knowledge so familiar to us today." He pays attention to the ingenuity and dexterity that allowed the relationship between weapon and organic activity to develop—how Gothic armament had become living machines. He adds, "It is rather strange that our times, which possess such extended knowledge of anatomy, cannot build a defensive clothing appropriate to our body" 39 —a perfect dovetail into the mechanized limb that concluded petit Jean's anatomy class.

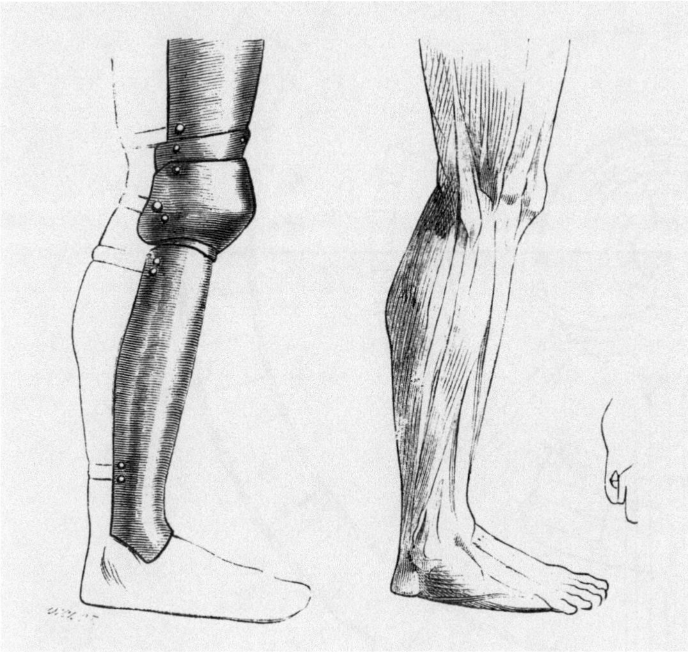

f.5 Eugène Emmanuel. Viollet-le-Duc. Fifteenth-century gauntlet. Steel engraving.

37 Viollet-le-Duc, *Dictionnaire raisonné du mobilier français*, vol. 5 (see note 33), p. 125.

38 Ibid., p. 484.

f.6 Eugène Emmanuel Viollet-le-Duc. Fourteenth-century armored shin-pad compared to the anatomy of the leg. Woodcut engraving.

39 Ibid., pp. 484–5.

What this last quote shows again is that, according to Viollet-le-Duc, the construction of armament, or any technical invention for that matter, is not so much an "application" of knowledge, as the product of a living process. Authentic invention condenses quasi-spontaneously disparate life circumstances: human needs shaped into the form of an object. It is in this regard, and in this regard only, that they become organic products whereby the boundary between the human and the non-human is eroded. With his study of medieval weaponry, Viollet-le-Duc establishes a filiation for French military tools, in the quasi-biological sense of the term. Showing how they were born out of the behavior of living Frenchmen. That is to say, of actions of the organism at work.

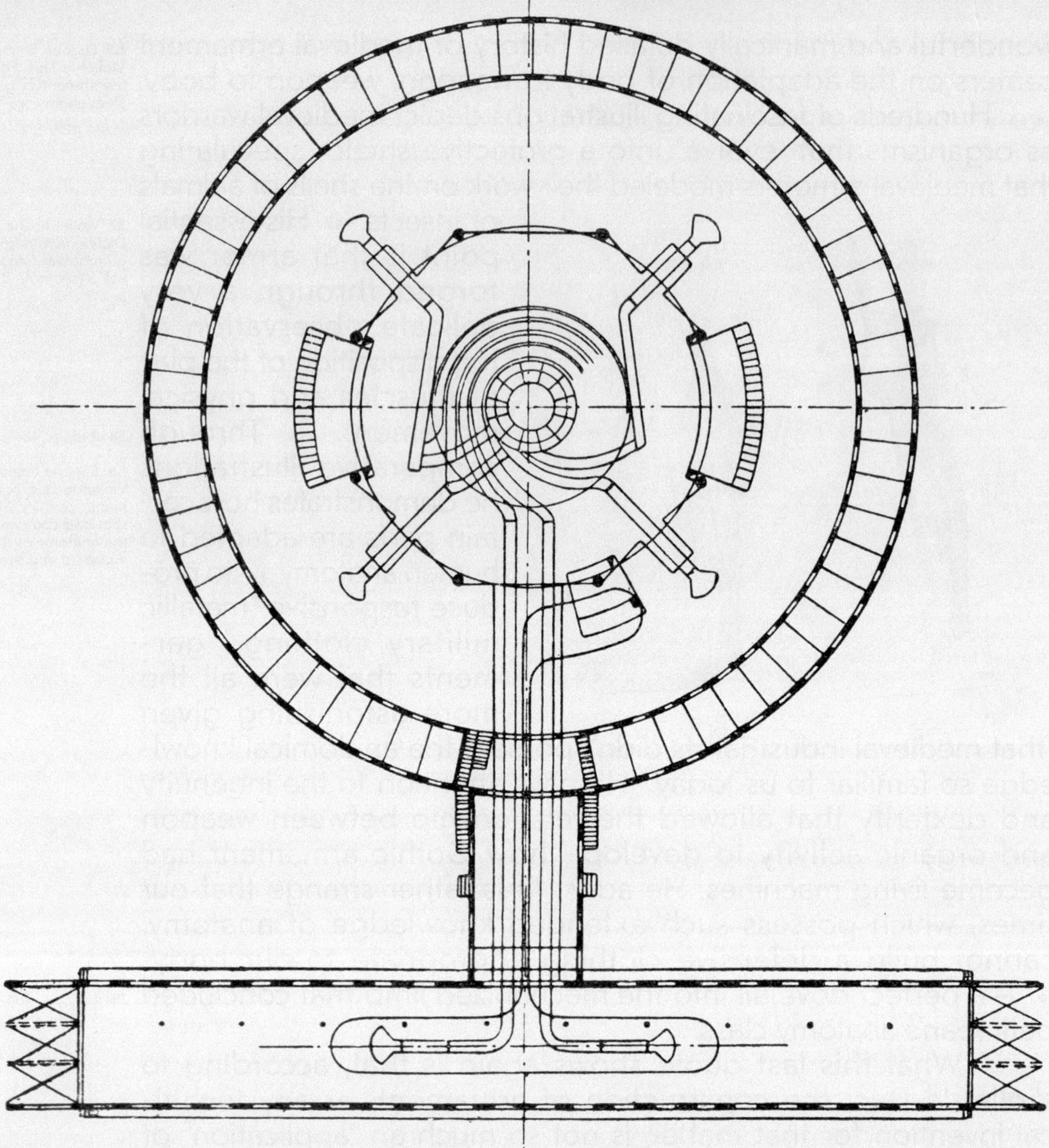

Abb. 13. Grundriß des Paketzustellamts in München.

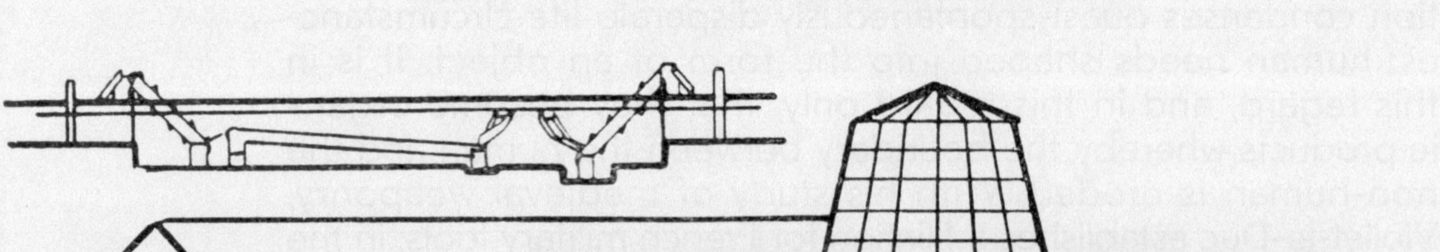

Abb. 14. Aufriß des Paketzustellamts in München.

Making Buildings Work
Susanne Jany

In 1926, a giant parcel-processing machine opened in central Munich—or at least this was how the newly built Paketzustellamt München-Marsfeld looked to its contemporaries. f.1 The distribution center was the first automated post office in Germany, with an extensive transportation infrastructure in place to execute major portions of its workflows. Parcels were collected from the nearby Munich central station by a special tramway then loaded on arrival onto electrically driven trolleys that transferred them to the sorting facility inside the central rotunda. f.2 The ground-level sorting unit exhibited 24 openings: four times six slots for the six main Munich postal districts, arranged in such a way that up to 12 employees could work simultaneously. Once the parcels were dropped through the slots, spiral chutes forwarded them onto

an electrically driven turntable that propelled them individually onto one of six conveyors—one for each postal district. f.3 The conveyors led to distribution tables at which postal workers sorted the parcels according to apartment blocks, streets, and house numbers, before stacking them along the outer wall of the building in special storage spaces equipped each with a door and a ramp. Postmen then accessed these spaces from outside, loaded the parcels into vans, and drove off to deliver them.

The distribution center specially designed and patented by a Berlin-based engineering firm, Mix & Genest, constituted a

Susanne Jany is Research Assistant at the Department of Cultural History and Theory at the Humboldt-Universität zu Berlin.

f.1 Paketzustellamt München-Marsfeld (1925–1926), floor plan and front elevation. The central sorting facility is located in the middle of the rotunda while its expansive conveyors extend throughout the basement.

f.2 Interior view of the central rotunda with electrically driven trolleys transporting parcels towards the "parcel distribution turbine."

closed system that conformed to the contemporary definition of "full mechanization": [1] its successive conveyors were motorized and ran independently of manual procedures. They were rigorously aligned in order to guarantee unhindered parcel processing, and operated from a central control panel situated at the supervisor's workstation in the rotunda. In contemporary reviews, machine imagery prevailed. Architect Walter Schmidt, who along with his colleagues Franz Holzhammer and Robert Vorhoelzer of the Munich postal planning department was in charge of the distribution center, called it a "giant conveyance apparatus," in

f.3 The sorting unit with its underground spiral chutes, turntable, and conveyors.

which "the operations, machine, and building are fused into a single entity." [2] Another member of the planning department, Oberpostrat Krinner, referred to the different elements of the complex as "construction components crucial to operations." [3] In particular the central sorting unit provoked machine references. It was repeatedly compared to technical gears and named a "parcel distribution turbine." [4] As Krinner explained, "like the guide vane that drives the water in a water turbine, it directs the parcels to their respective places." [5] Hans Schwaighofer, a leading expert in the mechanization of post offices, claimed that the "distribution turbine with its six conveyor systems constitutes a closed mechanism." [6]

Explaining a technical infrastructure in mechanical terms may well seem tautological and self-evident. However, I would like to argue that it was not only the electrically driven machinery

1 Cf. Hans Schwaighofer, *Post-Betriebsmechanik: Das Förderwesen in neuzeitlichen Postbetrieben grösserer Städte*, vol. 1: *Grundlagen der Post-Fördertechnik* (Wittenberg: Ziemsen, 1927), p. 69.

2 Walther Schmidt, *Amtsbauten: Aus Betriebsvorgängen gestaltet, dargestellt am Beispiel der bayerischen Postbauten* (Ravensburg: Otto Meier, 1949), p. 133. This and all further translations by Jill Denton.

3 "Das neue Münchener Paketzustellamt," *Archiv für Post und Telegraphie*, 55, no. 7 (1927), pp. 173–81; here p. 173.

4 Ibid., p. 179.

5 Ibid., p. 176.

6 Hans Schwaighofer, *Post-Betriebsmechanik*, vol. 2: *Mechanisierungsbeispiele aus der Postpraxis der Deutschen Reichspost-Verwaltung und des Auslandes* (Wittenberg: Ziemsen, 1927), p. 59.

of the Munich distribution center that prompted these various technical images. Around 1900, using machine metaphors to illustrate how certain buildings "work" was a means to explain not so much how to set things in motion mechanically but how to do so architecturally. The referential context was not merely the materiality of contemporary mechanized facilities—their transmissions, gears, rods, and so forth—but rather their disposition as operating entities. In light of this, the use of machine images is indicative of a new way of seeing architecture, one that emerged in the late nineteenth century and was characterized by a distinct awareness of the interrelation, in functional buildings, of work processes and their spatial prerequisites. Machine metaphors enabled building practitioners to envision architectural facilities as efficient operational systems and hence also to plan and build them as such.

1 In fact, machine imagery had been used to describe certain buildings years before the advent of systematic mechanization—also in regard to architectures seemingly unconnected with the mechanical realm. In a periodical from 1903, the Swiss surgeon Rudolf Ulrich Krönlein described the architectural disposition of the newly built Surgical Clinic at the Kantonsspital in Zurich in regard to its purposes, namely treatment and teaching:

"[I]t was to be hoped that the clinician would have an opportunity during the course of his training to experience the bustle and operations [Getriebe und Mechanismus] *of the aseptic surgical ward at first-hand and thus to witness the meticulous care and pedantry with which the surgeon endeavors before, during, and after an operation to meet the strictest demands of asepsis."* [7]

In similar fashion, the director of the surgical hospital in Tübingen, Georg Perthes, outlined the ideal layout of a surgical ward and suggested that "[t]he waiting room for sick persons, if procurable, should be separated from the operating room by several other rooms so that the waiting patients are not disturbed by their view of operations [*Getriebe*] in the surgical room." [8] The semantics of the German word Perthes chose here, *Getriebe,* oscillate between its more technical sense of "mechanical gears" and its everyday sense of *Treiben* and *Betriebsamkeit,* meaning "hustle and bustle" or simply "activity." The use of the term *Getriebe* links operating machines with certain actions, accomplishments, and operations in a specially designed functional building. In this perspective, a patient's proper treatment is acknowledged to be an architectural task. For, according to Perthes, disturbing elements

7 Rudolf Ulrich Krönlein, "Die aseptischen Operationsräume der Züricher chirurgischen Klinik und ihre Bedeutung für den chirurgisch-klinischen Unterricht," *Beiträge zur klinischen Chirurgie,* 37 (1903), pp. 660–75; here p. 672.

8 Georg Perthes, "Operationsräume," in Julius Grober (ed.), *Das deutsche Krankenhaus: Handbuch für Bau, Einrichtung und Betrieb der Krankenanstalten* (Jena: Fischer, 1911), pp. 526–56; here p. 527.

can easily be eliminated and the overall workflow improved by adding a few rooms and a couple of doors.

Thirty years before Krönlein and Perthes, the Swiss architect Eduard Guyer had published a book on hotel planning. Already in the first edition from 1874, his main concern was to emphasize the close relationship between the daily business of a hotel and its spatial disposition. [9] In the revised second edition from 1885, Guyer introduced the machine metaphor to clarify his point:

"To intertwine all [the various hotel operations], to arrange everything according to the scope of the business and in a purposive and clear way, to separate it from the guests' circulation and still come up with effective connections: the construction of this hospitality machine constitutes the most difficult task facing the architect and hotelier when building and arranging a hotel." [10]

In this perspective, the architectural layout appears to be the outcome of a careful spatial arrangement of different workflows—ranging from the delivery of food and supplies to the handling of luggage and the hotel administration. It is evident, therefore, that the notion of architecture as a machine is not contingent on specific technologies. A post office from as early as 1881 was able to be described as a machine-like structure without any need of motorized conveyors or mechanical turntables:

"The disposition of the operations building [Betriebs-Gebäude] is determined by the mechanism of the brisk and complicated dispatch operation. The prerequisites of its purposive disposition are ... generally little known and nor can they be overseen and evaluated from the counters." [11]

Since it was impossible to instantly assess this interrelation of work processes and their spatial requirements, considerable effort was put into investigating it: the German-speaking architectural community comprised of architects, engineers, building contractors, and operations experts published extensively on the matter in the late nineteenth century. The reference book *Handbuch der Architektur* and likewise architectural journals such as *Zeitschrift für Bauwesen*, *Deutsche Bauzeitung*, and *Zentralblatt der Bauverwaltung* were the central sites both for discussion of the "purposive disposition" (*zweckmässige Anlage*)—the prevailing discourse at the time—and proposed exemplary layouts of different building types. This also fostered a specific sensitivity to the interdependence of operations and spatial dispositions in functional buildings. In the decade 1870 to 1880, a nascent operative understanding of architecture was forged that I intend to address here as "process architectures," [12] my dual focus being how architectural structures are determined by workflows

9 Cf. Eduard Guyer, *Das Hotelwesen der Gegenwart* (Zurich: Orell Füssli, 1874), pp. 48–117.

10 Eduard Guyer, *Das Hotelwesen der Gegenwart*, 2nd, revised and expanded edition (Zurich: Orell Füssli, 1885), p. 86.

11 "Das Bauwesen der deutschen Reichs-Post- und Telegraphen-Verwaltung," *Deutsche Bauzeitung*, 15, no. 27 (1881), pp. 169–70; here p. 170.

12 Susanne Jany, "Postalische Prozess-architekturen: Die Organisation des Postdienstes im Medium der Architektur," *Archiv für Mediengeschichte*, 13 (2013), pp. 135–45. For a different historical concept of operative architectures, see Moritz Gleich, "Vom Speichern zum Übertragen: Architektur und die Kommunikation der Wärme," *Zeitschrift für Medienwissenschaft*, 12, no. 1 (2015), pp. 19–32.

and how workflows are determined architecturally. Discourse in the late nineteenth century about how to construct buildings as process architectures may be considered as the central historical site from which a logistic concept of functional architectures first emerged. Implicit in this concept, then broadly applied within the industrial, public, and private sectors, is an understanding of architecture as a means to organize and optimize workflows — as a *dispositif* (or apparatus) comprising architectural elements, circulating objects, technical infrastructures, discourses, operating procedures, and regulations, and designed with the criterion of purposive disposition in mind. 13 Whether the spatial organization of a factory, hotel, surgical clinic, or post office is concerned, the underlying assumption is that the production of commodities, the handling of clients, the treatment of patients, or the distribution of parcels can be structured and put into operation by architectural arrangements.

2 One of the basic spatial strategies for implementing workflows architecturally is to arrange workstations within a building in a way that corresponds to the logic underpinning the overall work process. 14 The movement of objects through the plant then coincides with their processing, and vice versa. In industrial laundry facilities, for instance, it is recommended that the requisite rooms and machines be arranged in a sequence that reproduces the "cycle that clothes are meant to undergo during laundry," 15 as shown by this exemplary floor plan for a midsize laundry facility, which was first published in *Der Gesundheits-Ingenieur*, a German journal on applied hygiene. f.4 Dirty clothes are received and sorted in the reception and sorting room before embarking on the laundry process: first via the

Fig. 132.

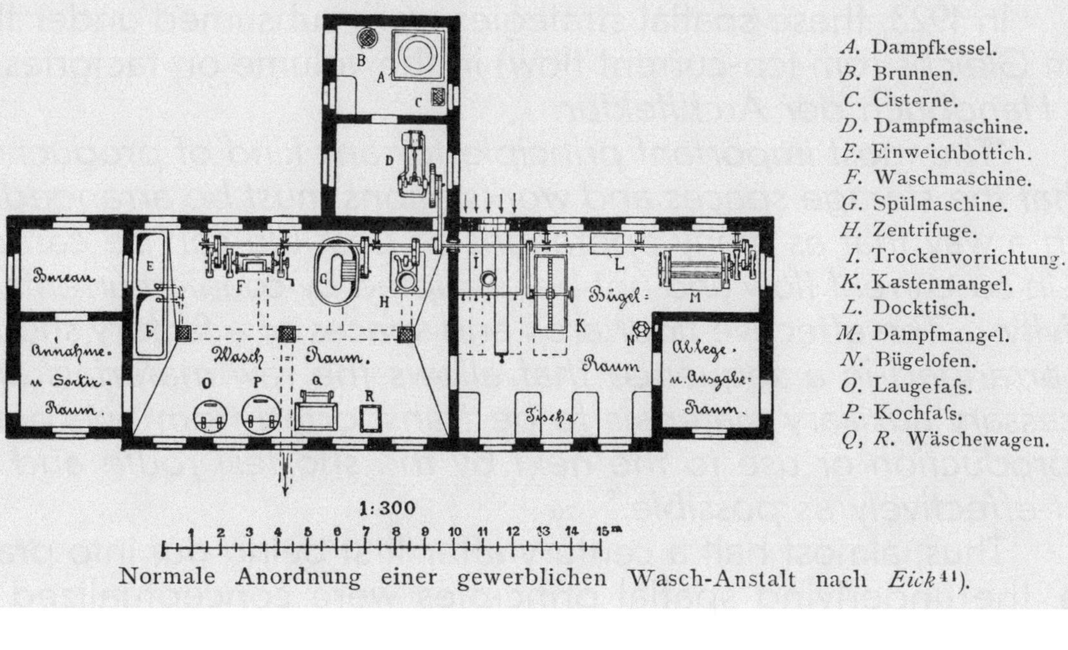

Normale Anordnung einer gewerblichen Wasch-Anstalt nach *Eick*[41]).

13 Over the last decade, a number of articles and anthologies have been published on logistic networks and infrastructures. The authors have investigated the material conditions of communication in terms of physical traffic, energy transmission, or social interaction. In exemplary case studies they have analyzed technical infrastructures embedded in domestic, industrial, or urban environments, such as automated production systems, container traffic, or the pneumatic post. But architecture too can adopt logistic qualities; especially when it is used as a technology to organize workflows. For examples from this infrastructure research, see Gabriele Schabacher, "Raum-Zeit-Regime: Logistikgeschichte als Wissenszirkulation zwischen Medien, Verkehr und Ökonomie," *Archiv für Mediengeschichte*, 8 (2008), pp. 135–48; Keller Easterling, "Disposition and Active Form," in Katrina Stoll and Scott Lloyd (eds.): *Infrastructure as Architecture: Designing Composite Networks* (Berlin: Jovis, 2010), pp. 96–9; Susanne Hauser, Christa Kamleithner, and Roland Meyer (eds.): *Architekturwissen: Grundlagentexte aus den Kulturwissenschaften*, vol. 2: *Zur Logistik des sozialen Raumes* (Bielefeld: transcript, 2013), pp. 14–21.

14 For alternative spatial strategies to organize workflows in functional architectures around 1900, see Susanne Jany, "Operative Räume: Prozessarchitekturen im späten 19. Jahrhundert," *Zeitschrift für Medienwissenschaft*, 12, no. 1 (2015), pp. 33–43.

15 Felix Genzmer, *Wasch- und Desinfektionsanstalten, Handbuch der Architektur*, part 4, vol. 5, no. 4 (Stuttgart: Bergsträsser, 1900), p. 83.

f.4 Exemplary floor plan of an industrial laundry.

washing room with its soaking tanks and washing machines, then via the pressing room with its drying ovens, mangles, and ironing boards. Once clean, the clothes are kept in the storage room until being returned to their owner in the reclaim room. The sequence of these four main rooms specifically reflects the order of their respective purposes and likewise the facilities and machines within them are conveniently set in close proximity to each other. This kind of spatio-logical arrangement was widely recommended at the time, inter alia for breweries, cotton mills, or slaughterhouses; [16] and equally, for hotels, railroad stations, post offices, or hospitals, which suggests that the architectural "handling" of people was to follow the same spatial rule. At railroad stations—due to the strict timeframe imposed by the train schedule—the traveler's route was supposed to lead directly from the ticket office to the baggage counter, then through the waiting room and barriers to the platform:

"The route and rooms through which the departing traveler passes must be arranged in a way such that he is advised or indeed compelled to take a certain path in order to reach the waiting room resp. the platform and train. Any futile seesawing or turning back of the traveler must be precluded." [17]

Unfavorable countercurrents in pedestrian traffic were also to be avoided: for instance, separate platforms for arriving and departing travelers were preferred in train stations, initially, because they were believed to prevent "countercurrents, which hinder the traveler's quick passage." [18] Under the primacy of workflows any architectural discrimination between circulating objects and subjects becomes obsolete. [19] The common spatial strategy is to accommodate short connections, a one-way direction, and a steady continuity while avoiding any form of collision, detour, crossway, or congestion.

In 1923, these spatial strategies were subsumed under the term *Gleichstrom* (co-current flow) in the volume on factories in the *Handbuch der Architektur*:

"The most important principle for any kind of production is that the storage spaces and workstations must be arranged in such a way that as many operations as possible can be carried out in co-current flow and not be disrupted by countercurrents of activity. ... The effective floor area and spaces of a factory should be arranged in a sequence that allows the raw materials and necessary auxiliary materials to be transported from one point of production or use to the next by the shortest route and as cost-effectively as possible." [20]

Thus, almost half a century after first being put into practice, the underlying spatial principles were conceptualized in

16 For instance, for breweries, see Friedrich Engel, et. al., *Landwirthschaftliche Gebäude und verwandte Anlagen ..., Handbuch der Architektur*, part 4, vol. 3 (Darmstadt: Diehl, 1884), p. 418.

17 Eduard Schmitt, *Empfangsgebäude der Bahnhöfe und Bahnsteigüberdachungen (Bahnsteighallen und -dächer), Handbuch der Architektur*, part 4, vol. 3, no. 4 (Leipzig: Gebhardt, 1911), p. 25.

18 R. Paulus, *Handbuch des Eisenbahnwesens in ökonomischer, rechtlicher, administrativer und technischer Beziehung*, vol. 2: *Bau und Ausrüstung der Eisenbahnen* (Julius Mauer: Stuttgart, 1872), p. 241.

19 This indifference had been characteristic of earlier conceptions of flows within buildings. Cf. Adrian Forty, "'Spatial Mechanics': Scientific Metaphors in Architecture," in Peter Galison and Emily Thompson (eds.), *The Architecture of Science* (Cambridge: MIT Press, 1999), pp. 213–31.

20 Wilhelm Franz, *Fabrikbauten, Handbuch der Architektur*, part 4, vol. 2, no. 5 (Leipzig: Gebhardt, 1923), p. 188.

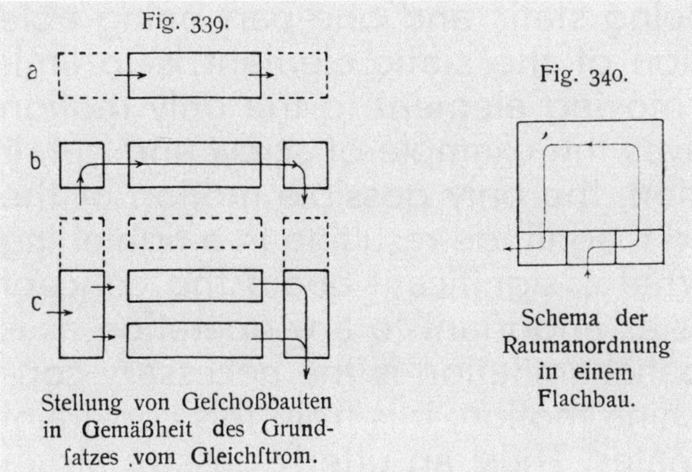

f.5 Layouts of cotton mills with spatial dispositions according to the *Gleichstrom* principle.

the meanwhile prominent terms of the rationalization discourse: direct flow lines, brief procedures, and cost reduction. Yet it was still basically about creating a continuous and steady flow of objects through a facility, in keeping with the operative logic of the overall production process. As is apparent from the schemes featured in the same *Handbuch* volume, spaces organized in co-current flow do not necessarily have to be in a linear sequence, as long as the movement of products or people within them is linear. **f.5** In fact, in most functional buildings, several workflows at once must be taken into account and architecturally structured. Accordingly, the task of the architect is to develop a spatial scheme that facilitates all the individual work processes in what is perceived to be the most advantageous way. To refer to mobilized elements as "currents"—whether as "co-current flow" in the 1920s or the aforementioned "counter-currents" [21] in the 1870s—not only helps bring this design problem to mind in order to be able to solve it, but also introduces a specific form of activity into the architectures concerned by featuring them as operative systems that initiate, direct, and regulate physical flows. Images of the fluid suspend the alleged contradiction between "static buildings" and "temporal processes," thereby opening up a specific concept of architecture that implies temporality, timing, and continuity while holding on to spatial matters of orientation, formation, and direction.

21 "Gegenströmungen," Paulus, *Handbuch des Eisenbahnwesens* (see note 18), p. 241.

3 What is the predominant concept of "machine" adapted within this specific field of architecture in the latter half of the nineteenth century? In particular, how is an operative (architecture) machine conceived? In 1875, the prominent German engineer and scientist Franz Reuleaux published a widely noted textbook on kinematics in which he systematically investigated machine motion. According to Reuleaux, a machine is "a combination of resistant bodies so arranged that by their means the mechanical forces of nature can be compelled to do work accompanied by certain determinate motions." [22] Machines can be put in operation because they consist of one or more mechanisms that are implemented as kinematic chains and transfer mechanical forces. These kinematic chains are made up of force-closed

22 Franz Reuleaux, *Kinematics of Machinery: Outlines of a Theory of Machines*. Trans. Alex B. W. Kennedy (London: Macmillan, 1876), p. 35.

element pairs, one part being static and one part being able to move. The main function of the static element is to limit the range of the actively moving element to the only motion requested. [23] Reuleaux gives the example of screw and nut. If the screw is fixed in position, the only possible motion of the nut is a helix of determined magnitude resulting in a tightening or loosening operation. What is significant about this concept is that the fixed element is as important to any operation as is the moveable element. Spatial limitation is the necessary condition for any form of machine motion; it is from this constraint that any mechanism originates. Thus, an utterly passive structure gains operative potential by enabling and directing the motion of an active element.

23 Cf. ibid.

This concept of mechanical operativity corresponds to how architects and building engineers conceived functional architectures around 1880. When they utilized spatial elements to accommodate workflows, guide movements, or trigger operations, "passive" structures such as walls, barriers, or ramps were understood to be parts of an operative system to the same degree that "active" elements such as circulations and workflows were incorporated as crucial parts of the architectural disposition. It was this alone that made operative elements of architectural elements: not because the latter needed to be moveable themselves but because they enforced and guided the motions of their complementary systemic elements. For instance, it is by constraining the streams of people or objects passing through that threshold elements gain the central control function ascribed to them in contemporary architectural media theory: "Openings like passages, entries, elevators, bridges, doors and windows are architectural elements with eminently operative character. They ... control the transport and transmission of persons, objects and information." [24] Accordingly, process architectures can be understood adequately only when workflows (or any other kind of flow) are taken into account as integral parts of the operative system. This is why architectural manuals such as the *Handbuch der Architektur* draw the architect's attention not solely to spatial arrangements but also to the logic of workflows, to architectural elements and technical equipment, to the streams of subjects or objects through and within a building, as well as to possible instances of interference or dysfunction.

24 Wolfgang Schäffner, "Architecture of the Openings: Windows, Doors and Switches," in Joachim Krausse and Stephan Pinkau (eds.), *Architecture of the Medial Spaces* (Dessau: Stiftung Bauhaus, 2006), pp. 74–9; here p. 78.

Reuleaux's ideas on the operating machine were not just descriptive in character. Rather, they investigated functional compositions in order to be able to optimize them: "[T]he technological part of the study of special machines," Reuleaux wrote in the 1870s, "examines ... by what special arrangement of the parts of the machine the required action can best be obtained." [25] The technologist tradition to which Reuleaux referred dates back to Johann Beckmann's *Anleitung zur Technologie* from 1777. In his treatise, Beckmann proposes the analytical description of handicrafts: classification of their elementary means and operations fosters an understanding of the manual production of objects, and is simultaneously the foundation for achieving technical and procedural innovation. Similar practices appear at the core of machine theory, for example in Jean Nicolas Pierre Hachette's *Traité élementaire des machines* (1811) or later in Robert Willis's *Principles of Mechanism* (1841). [26] Here, the analysis and construction of machines is based on the division, formalization, and optimized recombination of basic machine motions. In a general perspective seeking similarities in elementary motions and procedures, operating machines emerge as structural arrangements of technical elements that follow a certain functional logic and can be composed accordingly. Nineteenth-century engineering has been described as the practice of organizing and reorganizing (spatial) structures. [27] It was within the construction sector that the analytical approach to production and industrial labor was narrowed down to inquiries into the architectural means of ideally organizing workflows and circulations. Articles in architectural encyclopedias of the late nineteenth century need to be seen within this technologist tradition. Similar to Reuleaux's *Kinematics of Machinery* and other textbooks on machine theory, the *Handbuch der Architektur* appears as a medium for the analytical description both of buildings and the work processes executed within them, which is a prerequisite of their effective operation under optimal conditions.

In contrast to earlier conceptions, nineteenth-century "machines" are characterized in terms of their functionality, which is seen to derive from the distinct disposition of their elements—and this approach allows the machine label to be applied to operative entities on a variety of scales. As examples of kinematic machines, Reuleaux mentions water mills, high-pressure steam machines, and the railroad. Nineteenth-century notions of the machine are not limited to apparatuses but include mechanical systems of architectural or even larger dimensions. Hence, the analogy works in both ways: architectures adapt machine-like features and machines adapt architectural features—both thereby blur their respective boundaries. To draw an analogy between the machine and the factory is a common topos during the nineteenth century. The central criteria for machines are not, as in previous centuries, intrinsic values such as perfection and utility, but economic ones: machines must be productive and efficient;

25 Reuleaux, *Kinematics of Machinery* (see note 22), pp. 38–9.

26 Cf. Wolfgang Schäffner, "Erfindungskunst: Johann Beckmann und die Technologie der Künste im 18. Jahrhundert," in Inge Baxmann, Michael Franz, and Wolfgang Schäffner (eds.), *Das Laokoon-Paradigma: Zeichenregime im 18. Jahrhundert* (Berlin: Akademie Verlag, 2000), pp. 418–38; here pp. 429–38.

27 Historically: cf. Antoine Picon, *French Architects and Engineers in the Age of Enlightenment* (Cambridge: Cambridge University Press, 1992), p. 105. Theoretically: cf. Gilbert Simondon, *Du mode d'existence des objets techniques* (Paris: Aubier Montaigne, 1969), p. 117.

they are crucial elements within the national system of production. [28] Not surprisingly, behind the notion of the "purposive disposition" of functional buildings lay an economic imperative. The term *Zweckmässigkeit* (purposiveness) was broadly used by the national building authority, which administered the construction of public buildings and was bound to *Wirtschaftlichkeit*, the responsible handling of public funds. [29] Adopted likewise for industrial and private building projects, purposiveness and efficiency became the benchmark for all kinds of architecturally regulated workflows and circulations. [30] Describing functional buildings in terms of the machine around 1880 had a specific economical ring to it, namely to design them to run as effectively and profitably as possible. At the same time the machine reference proved to be the genuine tool to achieve just that: to analyze and optimize the architectures concerned with respect to their purposive disposition.

4 Imagining functional buildings such as banks, disinfection plants, or dairies as operating machines made up of intermeshing elements aids the understanding and implementation of workflows in relation to their spatial requirements and functional logic. Whereas the machine metaphor used in late-nineteenth-century architectural discourse only implicitly referred to matters of spatial organization, this aspect was increasingly explicated in the early twentieth century. In 1930, the engineer Rudolf Wolters wrote about railway stations:

"The most important elements of the layout occur for the same purpose [Zweckbestimmung] *and in the same order in each facility, just as every steam engine needs a boiler, cylinder, piston, and flywheel in this particular succession. Changing the sequence of the single components or omitting any one of them would render the machine impracticable. The station building can be compared to such a machine."* [31]

Being alert to the spatial interdependency of architectural and technical elements seems to have proven useful for avoiding scenarios such as that painted in vivid colors by Walther Schmidt, one of the architects of the Munich parcel distribution center:

"If from the very start the architect building a house pays no attention to the complex 'organism of the mechanical' then the technology thus ousted from the ... overall architectural organization of the project runs riot within the building. Then, there is no taming it. Then, spaces are blocked and windows obstructed. In every corner something judders, something shifts, and people are obliged to find their way around an ugly mess of a structure." [32]

28 Cf. Thomas Brandstetter, *Kräfte messen: Die Maschine von Marly und die Kultur der Technik 1680–1840* (Berlin: Kadmos, 2008), pp. 13–4, 227, 255–6.

29 For the novelty of this economic concept and its ties to architectural practice, see Reinhart Strecke, "Prediger, Mathematiker und Architekten: Die Anfänge der preußischen Bauverwaltung und die Verwissenschaftlichung des Bauwesens," in Strecke (ed.), *Mathematisches Calcul und Sinn für Ästhetik: Die preußische Bauverwaltung 1770–1848* (Berlin: Duncker & Humblot, 2000), pp. 25–7.

30 To give just one example: "In many cases the successful prospering of an industrial company depends on the purposeful arrangement [of its workspaces]." Julius Koch, "Mittheilungen über Fabriks-Anlagen," *Zeitschrift des österreichischen Ingenieur- und Architekten-Vereins*, 34, no. 4 (1882), pp. 74–81; here p. 74.

31 Rudolf Wolters, *Vom Grundriss der Empfangsgebäude großer Fernbahnhöfe* (Berlin: Germania, 1930), p. 41.

32 Schmidt, *Amtsbauten* (see note 2), p. 133.

In 1927, Krinner wrote of the Munich parcel center:

"[T]he design of the facility derives from operational requirements that were constantly borne in mind in their entirety. In consentaneous cooperation, the management, building constructors, and engineers have sought to find a solution that is strictly related to the facility's work processes." [33]

Evidently, not merely the parcel center's mechanical equipment was essential to its design but also the organization of its workflows. In fact, any mechanization or automation is based in the first place on the question of how to spatially accommodate the underlying work processes. The spatial organization of workflows precedes their mechanization. [34] The *Gleichstrom* principle is the *sine qua non* of what has come to be known as "assembly line production" or "flow production": [35] a successive order of workstations that reproduces the logic of the overall workflow and features short transportation distances and a preferably uninterrupted one-way flow of objects. The basic design principles of the Munich parcel distribution center thus date back to the way functional buildings were conceived and architecturally organized during the last third of the nineteenth century, namely as a means to make operational sense of their spatial arrangements. One of the most extensive descriptions of the parcel center can be found in a book written for architects and construction engineers under the telling title *Post-Betriebsmechanik*—"the mechanics of postal operations." [36] The title refers to the architectural implementation of technical infrastructures in post offices, which is outlined in the book. But at the same time, it suggests that there are certain mechanisms to postal operations themselves that need only be transferred to the architectural realm.

33 Krinner, "Münchener Paketzustellamt" (see note 3), p. 173.

34 See Gilbert Simondon's argument: "[C]e n'est pas le travail à la chaîne qui produit la standardisation, mais la standardisation intrinsèque qui permet au travail à la chaîne d'exister." Simondon, *Du mode d'existence* (see note 27), p. 24.

35 That is probably why the principle is referred to in 1923 as "co-current flow" and not as "assembly line production." In the *Handbuch der Architektur* it is presented as a spatial principle for the organization of workstations, machines, and transportation infrastructures that could be mechanized whenever possible but did not have to be: "Insistence on co-current flow is all the more important the larger the weight and amount of the raw materials and products. If the co-current flow cannot be implemented down to the last detail, fail-safe and cost-effective conveyor systems or means of transportation can offset the subsequent drawbacks." Franz, *Fabrikbauten* (see note 20), p. 188.

36 Schwaighofer, *Post-Betriebsmechanik* (see note 6), pp. 53–81.

Wires, Wirelessness, and the Morality of Form in "The New Machine Art"

John Harwood

John Harwood is Associate Professor at the Daniels Faculty of Architecture, Landscape and Design at the University of Toronto.

Note: Portions of this text have been published as parts of an article in *Media-N*; I thank the editors for their permission to republish certain turns of phrase in this new context. I also wish to thank Laurent Stalder and Moritz Gleich for inviting me to participate in a very stimulating conference, and Nicole Starosielski, Reinhold Martin, the media studies faculty reading group at Columbia University, and the Aggregate group for their constructive criticism on several versions of the text.

I concluded my last book, as so many historians and novelists do, with a discussion of a tangled skein. [1] There, I closed my attempt at a critical history of the design program at IBM in the post-Second World War era by remarking on the inability to date of art and architectural history to cope with the very language of technics and capital in the twentieth century. The crux of the problem remains to identify what, exactly, pits the technical against the aesthetic, the medium or apparatus against the work (a roomy set, that may be said to contain architecture as one of its many objects). Amongst other points of reference, I turned to the effort by Arthur Drexler to update the aggressively neo-Platonist and Thomist theses of Alfred H. Barr Jr. and Philip Johnson's 1934 exhibition at the Museum of Modern Art (MoMA), titled *Machine Art*, [f.1] with the rather lamely titled final entry, "The New Machine Art," in the museum's catalog of design objects of 1959. Whereas the 1934 exhibition had sought to stabilize the products of industrial civilization in aesthetic terms, claiming vociferously (if also anxiously) that a turbine, a lamp, a ball bearing, or a structural cable all deserved the status of object and artwork, [2] Drexler, illustrating his musings with the museum's recent and extraordinary acquisition of an IBM RAMAC control panel, reflected on what he saw as a fundamental redefinition of the status of the object in the face of the electrification of technics. [f.2] Then, as now, it is worth quoting him almost in full, precisely because

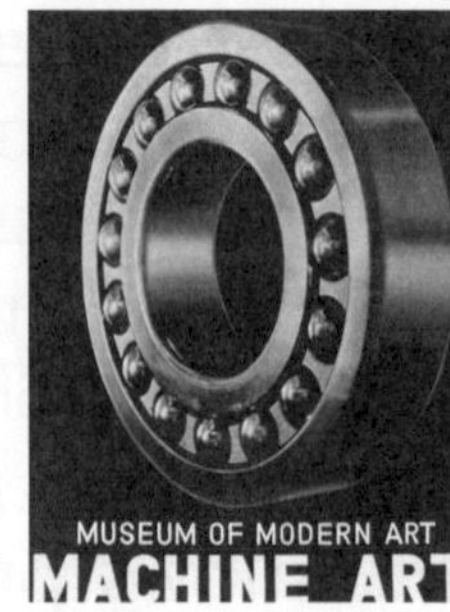

f.1 Alfred H. Barr Jr. and Philip Johnson, *Machine Art*, 1934.

f.2 IBM RAMAC control panel, 1958.

1 John Harwood, *The Interface: IBM and the Transformation of Corporate Design, 1945–1976* (Minneapolis: University of Minnesota Press, 2011), pp. 224–7.

2 Alfred H. Barr Jr. and Philip Johnson, *Machine Art* (New York: Museum of Modern Art, 1934).

Drexler both admitted and denied the existence of an ontological horizon beyond which the very grammar and syntax of art history and criticism would no longer be admissible. As he wrote on the final page of the catalog,

"Most often the design of machines is not consciously guided by aesthetic considerations. But in technology as a science, the more limited aesthetic decisions may be, the more significant are their effects. For this reason Machine Art still offers important clues to emerging concepts of design, the word design being understood not only in its conventional sense but also as a broad approach to the making and organizing of objects.

Since the end of World War II electronics has altered our conception of how things need to be shaped in order to work, and of how they may be related to each other. ... [T]he new machines are incomprehensible unless one knows about the existence of invisible forces."

So far, so good. The tangle of wires that seemed to swarm over the surface of the panel, which was detached from the rest of the apparatus it controlled, screwed to the wall of the museum, and photographed behind museum glass, could be seen as visible effects of the otherwise "invisible forces" that animated the data processor. Yet what becomes clear all too quickly is that the famed architectural critic, historian, and curator is only prepared to confront this technical artifact as an image—he is not only forced to confine his interpretation of that image as a representation of technical fact, but also as a representation and confirmation of Barr and Johnson's earlier hypothesis. Drexler again:

"Perhaps the most striking characteristic of the new machine aesthetic is its dematerialization of finite shapes into diagrammatic relationships. Examples are the printed electrical circuits, which replace separate three-dimensional objects with groups of patterns printed on a flat surface. Such patterns can hardly be said to have precise boundaries, or to be complete in themselves. This is also evident in a three-dimensional design such as the control panel from a RAMAC computer, with its clusters of colored wires arranged on a panel according to the requirements of computer operations.

... Dematerialization and pattern relationships recall similar ideas in painting, most notably in the work of Jackson Pollock." [3]

This unconvincing and even embarrassing recollection, I would like to argue, is evidence of a persistent problem haunting an art and architectural historical understanding of technics. But rather than review the subsequent iterations of this refusal to think through the RAMAC panel (or any other technical

3 Arthur Drexler and Greta Daniel, *Introduction to Twentieth Century Design from the Collection of the Museum of Modern Art* (Garden City, NY: Doubleday, 1959), p. 94. My emphasis.

artifact) as anything other than an image/object after MoMA's effort to domesticate all technical artifacts as museum objects—and moreover as objects that secure the boundaries of a number of subjects who regard them from a safe distance, but with growing anxiety—I'd like to take up an opportunity that I could not in my book, which is to try to address exactly why Drexler cannot cope with the wires. What erupts into the image in the attempt to reduce the RAMAC panel to an image is the visible, not invisible—and material as opposed to immaterial or "dematerialized"—presence of the connective tissue of the technical artifact. The RAMAC cannot be understood, under these conditions, as an object. The object, considered as a whole, even as a metonymic index to the real, requires different conditions. Following Franz Reuleaux and others, we should rather address the computer or portions thereof as an apparatus composed not only of objects but the crucial "linkages" between those objects that set the machinery in motion, [4] in which the literal wiring both produces objective effects and evades easy objectification. By contrast, what is required by the terms of modernist aesthetic ontology is the invisibility of the wires. We continue to live with this gap in our languages.

Drexler's perplexity was wholly in step with the erasure of wiring that had been the modus operandi of MoMA for decades; this requisite invisibility was forcefully theorized in the 1934 *Machine Art* exhibit and its catalog. As scholars such as Jennifer Jane Marshall have pointed out, Philip Johnson's careful staging of the exhibit, with all of its buffing and polishing, sought to emphasize the Phileban qualities of the mechanical and electrical products on display. [5] This reductive approach to aestheticizing industrial products as objects viewable by a stable subject—who, both Barr and Johnson in their short introductory essays stated with no shortage of drama, would be threatened by encountering many of these products in their original contexts—was also accomplished by literally detaching these products from the linkages to power and communications apparatus. [6] The appliances such as stoves and electric lights were displayed, both in the museum and catalog, with their power cables removed. When, in the first section of the exhibition, cables and wires were displayed, they were first of all selected carefully by Barr and Johnson for their mechanical properties and uses rather than their communicative properties, and were displayed as individual components rather than as relational elements within a larger structure or apparatus. Nowhere is this more evident than in the decision to display a sliver-thin section of a 3.5" steel wire rope manufactured by American Steel & Wire Co. [f.3] This crypto-orthogonal representation of the properties of the wire does not emphasize

4 Franz Reuleaux, *Kinematics of Machinery: Outlines of a Theory of Machines*. Trans. Alex B. W. Kennedy (London: Macmillan, 1876).

5 On the philosophical aesthetics motivating Barr and Johnson and on the display and photographic techniques used to "simplify" the industrial products on show in the exhibition, see Jennifer Jane Marshall, *Machine Art, 1934* (Chicago: University of Chicago Press, 2012), esp. chap. 3.

6 The epigraphs of the catalog express clearly the stakes of rendering the industrial machine and its products "beautiful." From Plato's *Philebus*, 51c: "By beauty of shapes I do not mean, as most people would suppose, the beauty of living figures or of pictures, but, to make my point clear, I mean straight lines and circles, and shapes, plane or solid, made from them by lathe, ruler and square. These are not, like other things, beautiful relatively, but always and absolutely." From St. Thomas Aquinas's *Summa Theologiae* I, q. 39: "For beauty three things are required. First, then, integrity or perfection: those things which are broken are bad for this very reason. And also a due proportion or harmony. And again clarity: whence those things which have a shining color are called beautiful." And finally, from L. P. Jacks's *Responsibility and Culture* (New Haven: Yale University Press, 1925), p. 62: "Industrial civilization must either find a means of ending the divorce between its industry and its 'culture' or perish."

the intertwining of elements so much as reduce the wire to a series of circular or cylindrical elements independent of one another, arranged together in a fixed pattern.

This reductivism has had a lasting impact on the way in which technical "objects"—and in the context of a museum, such artifacts or products are always understood as objects—are excluded from analysis under the terms of art historical method unless they may be construed as belonging to a canonical "medium" such as painting, sculpture, or printing, and even then these tools are examined primarily for evidence regarding the hand and creative mind that wield them. The technical artifact—whether construed as object (tool), subject (automaton), element of a larger apparatus (organ), medium, or *dispositif*—unsettles and disrupts a neo-Platonic, even neo-Kantian desire to disaggregate the world into subjects and objects; and it does so in a way that provokes ethical and moral objections to liberalism and challenges the grammar and syntax of art history. 7 So, rather than move forwards from "Machine Art" and trace a historical legacy from that fundamentally conservative effort to tame the technical artifact, I propose moving backwards, genealogically. As Bernhard Siegert has shown in his in-depth study of the origins of the modern postal system, and in his later essays on the complex arrays of *Kulturtechniken* deployed in articulating other modes of telecommunication, infrastructures carry "meaning" from place to place, replicating themselves as they go. 8 It is in this way that, as Friedrich Kittler so provocatively stated, "media determine our situation." 9 Yet it is the specificity of the wire—its configuration as a particular form of network, carrying particular information—that matters most in this open-ended determinism. As Florian Sprenger noted in his recent studies of "immediate media," the cables and wires and other apparatus of electrical and electronic media are difficult and nearly always invisible as either subjects or objects; they are "in-betweens." 10 Yet what are these wires, that they must be removed from the image, from the analytical understanding of objects as understood by the history of art and architecture?

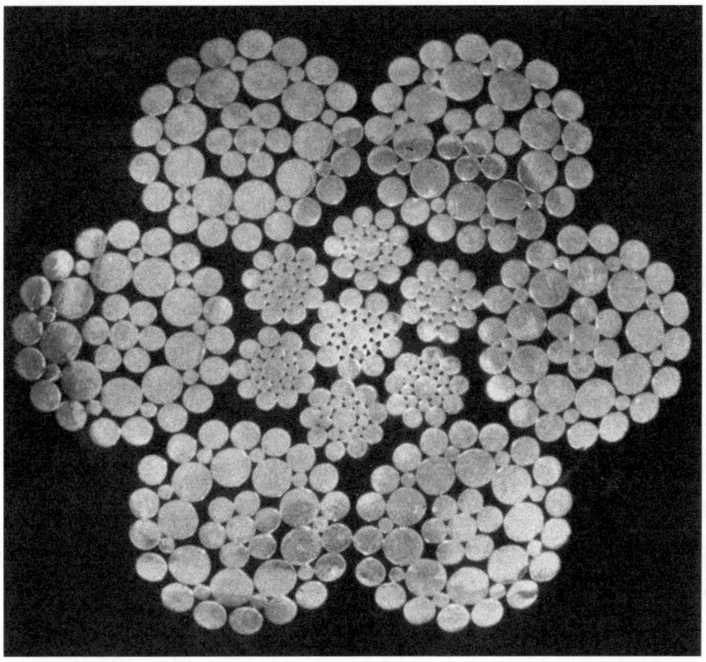

f.3 American Steel & Wire Co. (Subsidiary of United States Steel Corp.), section of wire rope, 3.5" in diameter, 1934.

7 Interestingly, in an effort to reestablish the methodological bases of art history in the wake of poststructuralist and postcolonialist critiques, Robert S. Nelson and Richard Shiff commissioned 22 essays on "critical terms." Neither "object" nor "subject" nor the term used to collapse one onto the other—"style"—are among the terms examined, although "style" was belatedly included in the revised edition of 2003. Amazingly, "medium" is also excluded. Nelson and Shiff (eds.), *Critical Terms for Art History* (Chicago: University of Chicago Press, 1996).

8 See Bernhard Siegert, *Relays: Literature as an Epoch of the Postal System*. Trans. Kevin Repp (Stanford, CA: Stanford University Press, 1999); and Siegert, *Cultural Techniques: Grids, Filters, Doors, and Other Articulations of the Real*. Trans. Geoffrey Winthrop-Young (New York: Fordham University Press, 2014).

9 The full statement from the preface to Friedrich Kittler, *Gramophone, Film, Typewriter*, trans. Geoffrey Winthrop-Young and Michael Wutz (Stanford: Stanford University Press, 1999 [1986]), p. xxxix, reads: "Media determine our situation, which—in spite or because of it—deserves a description."

10 See Florian Sprenger, "Between the Ends of a Wire: Electricity, Instantaneity and the Globe of Telegraphy," in Michaela Hampf and Simone Müller-Pohl (eds.), *Global Communication Electric: Actors of a Globalizing World* (Frankfurt: Campus, 2013), pp. 355–81; and Sprenger, *Medien des Immediaten: Elektrizität, Telegraphie, McLuhan* (Berlin: Kadmos, 2012).

A good place to start pulling on the end of the wire is with the beginnings of the usage of wires as key components of telecommunications media. As the first scientifically (as opposed to alchemically) minded commentator on the role of metals in modern culture, Georgius Agricola wrote in defense of the mining and metallurgical arts of "the arguments which may be used against this art, and against metals and the mines:" [11]

"If we remove metals from the service of man, all methods of protecting and sustaining health and more carefully preserving the course of life are done away with. If there were no metals, men would pass a horrible and wretched existence in the midst of wild beasts; they would return to the acorns and fruits and berries of the forest. They would feed upon the herbs and roots which they plucked up with their nails. They would dig out caves in which to lie down at night, and by day they would rove in the woods and plains at random like beasts, and inasmuch as this condition is utterly unworthy of humanity, with its splendid and glorious natural endowment, will anyone be so foolish or obstinate as not to allow that metals are necessary for food and clothing and that they tend to preserve life?" [12]

The argument is plain enough. The sophistries of those who would object to mining and metallurgy or manufacture and medicine risk plunging contemporary society into chaos. But the argument also claims that metals actually are constitutive of the relations between people in a civilized society, and this is both more important to Agricola's project to recuperate and perfect the sciences and techniques of mining and metallurgy, and more relevant to the pocket history of wires that follows. Agricola depicts a metallurgical society that is shot through with metal artifacts which make possible greater specialization of labor, cultural variation, advances in medicine and other sciences, and so on. The bonds that knit the "beasts" of the wild into the "humanity" Agricola identifies in his sixteenth-century present, however, are wireless. First used for making basic tools, then for making jewelry, then for making money, then for more advanced machinery (including weapons) — throughout history, Agricola reminds us, metals were constitutive of community and communications: they allowed ever more sophisticated and intimate cultural relationships to act over distance. In other words, metals are always already media.

As the first "short history" of telephony written after the invention of the Bell system relates, the "ear trumpet" and "speaking trumpet" were early instances of communicating sound at great distance, allowing sound waves to travel further through the medium of air because "the walls of the tube reduced the loss of

11 Georgius Agricola, "Preface," in Agricola, *De Re Metallica*. Trans. Herbert Clark Hoover and Lou Henry Hoover (New York: Dover, 1950), p. xxix.

12 Ibid., p. 14.

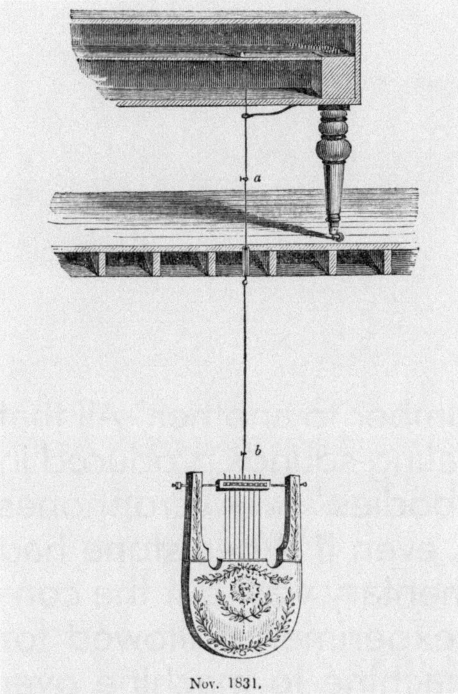

energy from spherical action" 13 —that is, the increase in volume (or prevention of decay) of a given sound was accomplished through tectonic means. Even though many inventors can lay claim to the discovery of a means of telecommunication through a solid (as opposed to gaseous) medium well beforehand, the scientist, inventor, and entrepreneur Sir Charles Wheatstone deserves credit for picking up an often ignored observation from the preface to Robert Hooke's 1665 *Micrographia*. As Hooke boasted there, "I can assure the reader that I have, by the help of a *distended wire*, propagated the sound to a very considerable distance in an *instant*, or with as seemingly quick a motion as that of light, at least incomparably swifter than that which at the same time was propagated through the Air; and this not only in a straight line, or direct, but in one bended in many angles." 14 In the late 1820s, Wheatstone resuscitated these experiments in a sequence that would lead him to share in developing the first commercial electrical telegraph, in parallel with his ongoing research into the nature of sound as propagated via vibrations in the air.

After reproducing Hooke's experiments and devising his own, Wheatstone demonstrated that a "conducting-wire for transmission of sounds" could be attached to the sounding board of a musical instrument such as a piano-forte or a harp at a perpendicular angle, which would then transmit the sounds to a similar instrument via its sounding board at the other end. f.4 Moreover,

"The sounds of an instrument may be at the same time transmitted to more than one place; ... it would be possible to extend a horizontal conductor through a series of rooms belonging to different houses, and (provided the instrument connected with one of its extremities be constantly played upon) to hear at pleasure the performance in any of these rooms, by merely attaching a reciprocating instrument to the conductor; on removing this instrument, the sonorous undulations would pass inaudibly to the next apartment." 15

With this set of observations, the wired world was born, a mass media challenge to the primacy of print and other visual media (such as the optical telegraph), and Wheatstone was one of its most aggressive theorists. He noted that because

13 John E. Kingsbury, *The Telephone and Telephone Exchanges: Their Invention and Their Development* (London: Longmans, Green & Co., 1915), p. 7.

f.4 Sir Charles Wheatstone, "Conducting-wire for the transmission of Sounds," engraving, 1831.

14 Robert Hooke, *Micrographia; Or, Some Physiological Descriptions of Minute Bodies Made by Magnifying Glasses with Observations and Inquiries Thereupon* (London: Jo. Martyn and Ja. Allestry, 1665), n.p. Emphasis in original.

15 Charles Wheatstone, "On the Transmission of Musical Sounds through Solid Linear Conductors, and on their Subsequent Reciprocation," *Journal of the Royal Institution of Great Britain*, 2 (1831), pp. 223–38; here pp. 232–3.

f.5 "Multiple Switchboard at Baltimore," engraving, 1884.

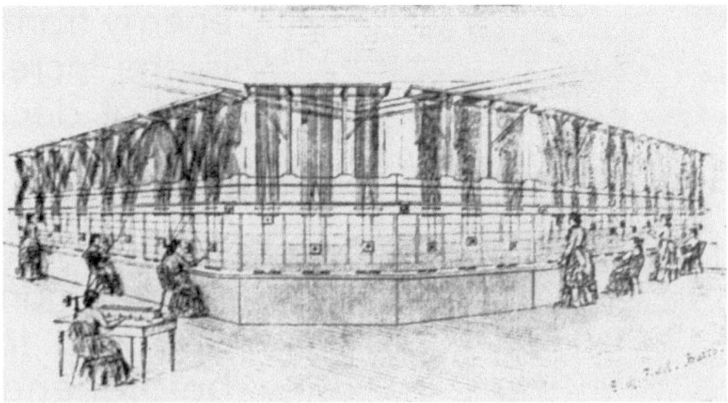

sound travels roughly 17 times faster through "iron wire, glass, cane, or deal-wood rods" than it does through the air, it would be "as easy to transmit sounds through such conductors from Aberdeen to London, as it is now to establish a communication from one chamber to another." All that was needed was a means of "communicating sounds produced in the air with sufficient intensity to solid bodies." [16] Microphones and amplifiers were some ways off yet, even if Wheatstone had developed the former device in a rudimentary way, but the conceptual organization of Wheatstone's experiments allowed for a full vision of communication from machine to machine over vast distances at high speeds.

16 Ibid., pp. 237–8.

The wired world, as even Wheatstone's interconnected musical instruments made clear, was one in which the relationship of architecture to telephony was inverted. Instead of serving as a direct means of bounding, reflecting, and projecting sound, architecture had to be penetrated by "solid media" of a different technical nature. A glance at a multiple switchboard from the 1880s is enough to carry the point for now: the intricate web of wires penetrates the architecture of the telephone exchange building according to its own logic, and requires the design of an interface at the scale of furniture in order to make it manageable for its female operators. f.5

In short, wires operated against the grain of the visual and spatial logic of architecture, and rearranged people and their activities. The tidy system of boundaries that Alberti prescribed to divide public from private, sacred from profane, and individual from family [17] —all of these would be called into question by a new method of delineation that would pass through architecture on a perpendicular axis. Nowhere is this more helpfully theorized than in one of the most famous texts on wiring, Adam Smith's *An Inquiry into the Nature and Causes of the Wealth of Nations*. At the very outset of that treatise on liberal political economy, Smith famously called the reader's attention to the modern manufacture of a "very trifling" kind of thing, "but one in which the division of labour has very often been taken notice of": the pin.

17 Leon Battista Alberti, *The Ten Books on Architecture: The 1755 Leoni Edition*. Trans. James Leoni (Mineola: Dover, 1986), p. 2.

"[I]n the way in which this business is now carried on, not only the whole work is a peculiar trade, but it is divided into a number of branches, of which the greater part are likewise peculiar trades. One man draws out the wire, another straights it, a third

cuts it, a fourth points it, a fifth grinds it at the top for receiving the head; to make the head requires two or three distinct operations; to put it on, is a peculiar business, to whiten the pins is another; it is even a trade by itself to put them into the paper; and the important business of making a pin is, in this manner, divided into about eighteen distinct operations, which, in some manufactories, are all performed by distinct hands, though in others the same man will sometimes perform two or three of them." [18]

Smith goes on, of course, to note that even a relatively limited degree of the separation of labor allowed these tradespeople, a factory of "ten persons," to produce 48,000 pins per day, when "they certainly could not each of them have made twenty, perhaps not one pin in a day." Therefore, Smith concludes proudly, "the effects of the division of labour are similar to what they are in this very trifling one." [19] Because of the division of labor, tradespeople became better at their more sharply defined tasks, time was saved, and machinery could be applied to the various piecework tasks. Q.E.D. What passes unnoticed by most readers of Smith, however, is that Smith's moralizing (remember that he and his university identified him as a moral philosopher, not a political economist) about efficiencies of labor, production, and exchange of manufactured goods is formulated through an obsessively woven and cut network of wires and threads. The pin may be a "trifle," but it is a very particular trifle, one worthy of a substantive entry in the *Encyclopédie* of Diderot and d'Alembert, from which Smith cribbed his facts. [20/f.6]

The irony that scholars have assumed dramatizes Smith's text is well known: look how simple the pin is; yet it requires such a complex aggregation of labor and technics to produce! But less often emphasized is another of Smith's conceptual and literary motives for narrating pin manufacture as the Trojan Horse of liberal capitalism. [21] The wire or its refined form as a pin is a connector, a *copula*: it holds things together in novel and temporary arrangements, as the linguistic or technical component that binds subjects to predicates in particular forms. [22] As such it provides a direct link to Smith's earlier writing in moral philosophy, in particular to his late addition to *The Theory of Moral Sentiments*: a "Dissertation on the Origin of Languages." There, Smith sought to bolster his arguments regarding the familiar concept of the "impartial spectator" by developing a theory about the distinction between ancient and modern languages. Ancient languages, per Smith, relied upon cases to define abstract relations between objects; modern languages, he argues, supersede ancient languages in one important way, which is that they supplant case with an increasingly vast array of prepositions:

18 Adam Smith, *An Inquiry into the Nature and Causes of the Wealth of Nations* (London: W. Strahan and T. Cadell, 1776), p. 6.

19 Ibid., p. 7.

20 "Épinglier," *Encyclopédie ou Dictionnaire raisonné des sciences, des arts et des métiers: Recueil des planches*, 3rd ed. (Paris: Briasson, David, Le Breton, 1765), pp. 1–8.

21 For a decisive rereading of Smith's writings on political economy — so often considered in fragmentary and isolated form — in relation to his and others' writings in moral philosophy, see Mike Hill and Warren Montag, *The Other Adam Smith* (Stanford: Stanford University Press, 2015).

22 On the linguistic and syntactical theory of *copulae*, see Andrea Moro, *The Raising of Predicates: Predicative Noun Phrases and the Theory of Clause Structure*. Cambridge Studies in Linguistics, 80 (Cambridge: Cambridge University Press, 1997), esp. chaps. 1 and 4.

f.6 "Épinglier," *Encyclopédie ou Dictionnaire raisonné des sciences, des arts et des métiers*, 1765. → 90/91

A
B
Fig 1.
Fig 15.
Fig. 21. 1
Fig. 16. N° 2
M
Fig 16.
Fig. 21.
Pieds

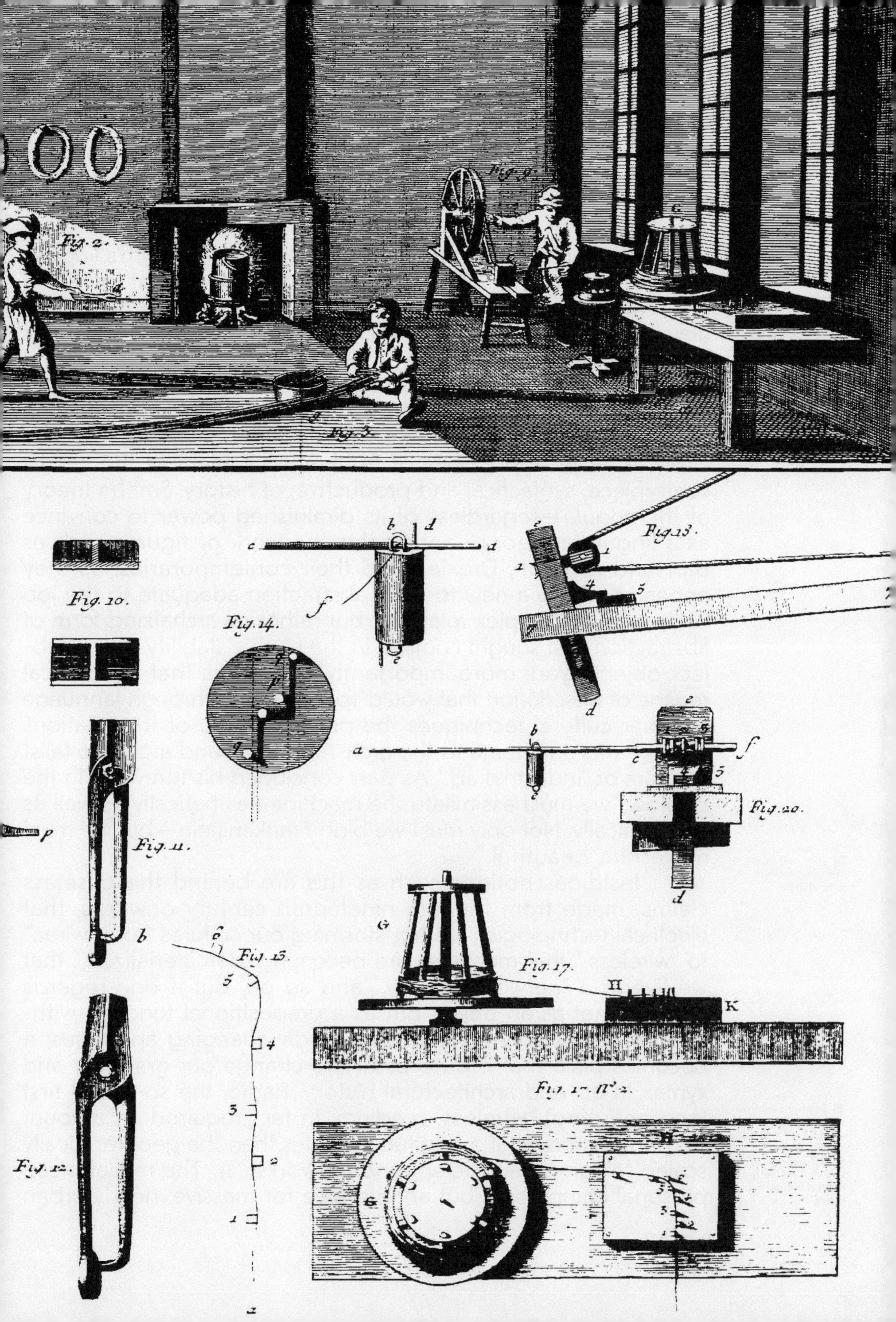

Fig. 2.
Fig. 3.
Fig. 9.
Fig. 10.
Fig. 11.
Fig. 12.
Fig. 13.
Fig. 14.
Fig. 15.
Fig. 17.
Fig. 17. N°. 2.
Fig. 20.

"A Preposition denotes a relation, and nothing but a relation. But before man could institute a word, which signified a relation, and nothing but a relation, they must have been able, in some measure, to consider this relation abstractedly from the related objects; since the idea of those objects does not, in any respect, enter into the signification of the Preposition. The invention of such a word, therefore, must have required a considerable degree of abstraction." [23]

Although I have no intention of embracing Smith's linguistic theory — certainly we cannot accept the notion that the novel invention of prepositions would be sufficient to the work of describing the complexity of modern technics — the value of his observations on language and his efforts to retheorize political economy as a matter of seemingly invisible interconnections is profound. What Smith asked his readers to rethink, in both the *Theory of Moral Sentiments* and the *Wealth of Nations*, was precisely the valedictory wave of the human subject as the crucial centerpiece, syntactical and productive, of history. Smith's theory of the *copula* — regardless of its diminished power to convince as a linguistic theory — anticipates the work of figures such as Barr and Johnson, Drexler, and their contemporaries, as they appealed not to a new form of abstraction adequate to the job of describing complex relations, but rather an archaizing form of abstraction that sought comfort in the mythic stability of the subject/object dyad; more importantly, it appears that a technical regime of description that would specify either through language or other cultural techniques the precise nature of the relations at stake was anathema to the arch-modernist and arch-capitalist curators of "industrial art." As Barr concluded his foreword in the catalog: "we must assimilate the machine aesthetically as well as economically. Not only must we bind Frankenstein — but we must make him beautiful." [24]

Insidious notions such as this are behind the baseless claims, made from the late nineteenth century onwards, that electrical technologies are transforming our cultures from "wired" to "wireless," that machines are becoming "dematerialized," that we live in a "network society," and so on. But if one regards the wire, not as an object but as a prepositional function within a very decidedly material but rapidly changing apparatus, it becomes plain that it is necessary to change our grammar and syntax in art and architectural history. Radio, the so-called first (and enduring) "wireless" medium, in fact required an amount of wiring an order of magnitude greater than the geographically scaled telegraph and telephone networks. [25] The miniaturized personal computer is but an interface for massive, nearly urban

23 Adam Smith, "Considerations Concerning the First Formation of Languages, and the Different Genius of original and compounded Languages," in Smith, *The Theory of Moral Sentiments: To Which Is Added a Dissertation on the Origin of Languages*, 3rd ed. with dissertation (London: A. Millar, A. Kincaid and J. Bell, 1767), pp. 437–78; here p. 449.

24 Barr and Johnson, *Machine Art* (see note 2), n.p.

25 For the best theoretical consideration of radio yet published, see Rudolf Arnheim, *Radio*. Trans. Margaret Ludwig and Herbert Read (London: Faber & Faber, 1936).

scale architectural enclosures filled with even more wires. And the wires themselves are the site of extraordinary conflicts and manipulations: whether discussing the viability of something called "neuroaesthetics" as a means for figuring out how we are "wired," or the creation of miles-long loops of wire to delay a competitor's efforts to purchase derivatives, what is certain is that there are people for whom much is at stake in preserving the invisibility of wires. [26]

26 The wire, in fact, is the "smoking gun" in a recent work of popular non-fiction: a small box containing 38 miles of wires surreptitiously installed by investment bankers in a rival's telecommunications apparatus in order to slow and interrupt his ability to trade in "real time." See Michael Lewis, *Flash Boys* (New York: W.W. Norton & Co., 2014). For an excellent scholarly analysis of the geopolitics of wiring and cables, see Nicole Starosielski, *The Undersea Network* (Durham: Duke University Press, 2015); and for a more theoretically driven examination of the "materiality" of the "medium of mediums" that "bind the world," see Daniel Gethmann and Florian Sprenger, *Die Enden des Kabels: Kleine Mediengeschichte der Übertragung* (Berlin: Kadmos, 2015).

The Form of the Cloud: Kersten Geers's Architecture Without Content

Caspar Schärer

The cloud ranks as the most flowery term yet to have emerged from the digital world order. Undoubtedly there are more menacing weather phenomena brewing on the horizon, but the cloud's connotations are altogether positive—an intangible nimbus in which we are encouraged to store all our data. We are, of course, fully aware that this is not the whole story, since the ostensibly ephemeral cloud also has a physical manifestation. It is a machine, consisting of thousands if not millions of machines, bundled and pooled in data centers of grotesque dimensions. Exabytes (10^{18} bytes) of data flow daily through these machines, where they are stored and then dispersed throughout the world again.

In a manner of speaking the cloud casts a shadow on the earth's surface. These Internet big boxes belonging to Apple, Microsoft, Facebook,and Amazon are scattered across the entire planet. They require vast sites, which is why they are located where land is cheap; they use up a lot of electricity, which is why the operators seek to place them near power plants; they have to be secure, which is why they are situated in remote parts of the globe, gathering in clusters in Sweden at the Arctic Circle, in Iceland, and in the US states of Nevada and

Oregon, where the data centers of Facebook and Apple stand on opposite sides of the same street on a plateau close to Prineville. This physically constructed Internet infrastructure is impressively large, and its organization highly efficient. It is, however, simultaneously largely hidden from view and so not perceived as such. Neither the standardized frame—the 19-inch rack for mounting the server and memory modules—nor their accumulation in data centers with dimensions of up to 500,000 square meters are anchored in popular consciousness as the cornerstones of the Internet. This is all the more remarkable when we consider that particularly on the Internet information is constantly being exchanged and disseminated on innumerable platforms: in other words, knowledge is being produced.

Machines that connect people in this manner should embody and emanate more architectural importance. This is the view of the Belgian architect Kersten Geers, who persistently tackles the topic of the big box in his design studios at Columbia University in New York, at the TU Graz, the Accademia di architettura in Mendrisio, and the EPF Lausanne. All the data centers, and likewise the logistical hubs, distribution centers, and waste treatment plants, form an enormous structural mass that arises bereft of any architectural detailing. Correspondingly, Geers has provocatively titled his seminar series "Architecture

Without Content" — the implication being that it must be possible to produce good architecture even minus a complex program and articulated spatial sequences. It may be that the content happens to not be irrelevant, according to Geers, but it lies beyond the architect's influence. Often the room program is quite simply banal, as for instance with an office building where the client precisely dictates what he or she wants executed. Working merely on the shell leaves many architects frustrated, or, conversely, they resort to designing demonstratively effusive façades. Kersten Geers takes a far more radical approach, encouraging his students to take a closer look and productively exploit the freedom this very lack of content provides.

The students investigate exemplars, so-called ancestors, such as Venturi and Rauch, Aldo Rossi or Andrea Palladio, in order to unearth their architectural quintessence. Armed with these proven principles and tools they work on their projects for a big box. By this means the architectural DNA of the ancestors becomes transferable to seemingly faceless infrastructure buildings, for instance data

centers. The basis of the stance that Geers imparts to his students is a cold and yet fascinating vision of the urbanized world. He refers to an "even-covered field," a uniform carpet of development, somehow city, and somehow nonetheless not. In this "field" the buildings stand differently than in the densely developed city. As a rule there is no front or back. Moreover, large structures have another significance: they act as moorings in the fabric of the unending agglomeration. For this specific reason they deserve more architectural diligence.

In this respect it is only logical that in some of the design semesters not even the urban-planning context is given—the only question is the building itself. At first glance this may seem irritating (Each project must respond to its context!), but on closer

consideration it is in fact perfectly consistent. On an "even-covered field" the immediate surroundings are sometimes really not that important. The big box is so large that it potentially defines the context itself, even when it has no public function. No content, no context, only architecture: together with his students, Kersten Geers delves to the core of the discipline that all too often lies forgotten beneath a tangled canopy of conflicting requirements. And it is precisely by freeing the disconnected big box from its isolation and by—with almost naive affection—maximally recharging it with architecture that he reestablishes its compatibility with a world where addresses have long since become virtual.

f.1/f.2 Kersten Geers, *Architecture Without Content*, vol. 4: *Necessary Architecture* (London: Bedford Press, 2015), n.p.
f.3/f.4 Ibid., vol. 5: *Places of Accumulation* (London: Bedford Press, 2015), n.p.

Vitaglass: A Modern Boundary Technology between Laboratory Research, Architecture, Public Health, and Environmentality in the 1920s and 1930s

Kijan Espahangizi

Kijan Espahangizi is Managing Director of the Center "History of Knowledge" (ZGW) at the University of Zurich and ETH Zurich.

f.1 Rachitic rats behind Vitaglass, 1928.

"A laboratory in the university, used by my department, has three windows facing south. The sashes of two of these windows were removed and replaced by two Vitaglass panes each. By building a shelf in front of these three sashes on which to place rat cages, and by using screens to cut out cross rays, it was possible to measure the preventive action of transmitted rays through each of four panes of Vitaglass and through window glass as recorded by the rat occupants of these cages on a basal diet of rickets-producing power (Sherman-Pappenheimer No. 84). Three 30-day tests were run." 1/f.1

This experimental setting was part of a research project at the Laboratory of Physiological Chemistry at Columbia University in New York in 1928. The goal was to determine the "antirachitic value" of a new kind of window glass, which had been given considerable public attention since the mid-1920s. This so-called Vitaglass—which even made it into *Time* 2 —was transmissible to the "healthiest" part of sunlight, ultraviolet rays. Long before

1 Walter Eddy, "The Use of Ultra-Violet Light Transmitting Windows," *American Journal for the Public Health and the Nation's Health*, 18, no. 12 (1928), pp. 1470–9; here p. 1474.

2 "New Glass," *Time*, November 1, 1926, and "Sun & Glass," *Time*, May 23, 1927.

the discovery of the carcinogenic effect that dominates our view of these rays today, they seemed to promise unambiguous vitality and health, as a cure for the disease of rickets, for example.

When Vitaglass hit the market, it provoked a public and scientific debate on ultraviolet rays and windows that lasted until the mid-1930s. [3] The product itself never met the initially high expectations. [4] Yet from an analytical point of view, this historical episode offers valuable insight into the interactions of architecture, laboratory research, and biopolitics in the early twentieth century, and also sheds light on the role of glass boundaries as a material medium of modern environmentality. I use the term environmentality to refer not to a certain political attitude towards ecological issues, but rather to a specific mode of reasoning and imagination based on careful appraisal of the conditions surrounding epistemic objects. [5]

The experimental setting at Columbia University, which was accompanied by parallel tests in a village schoolroom in the vicinity of New York City, illustrates the convergence in the 1920s of architectural, health-enhancing, and experimental environments, whose respective boundaries coincided in the glass pane. In order to determine whether exposure to a certain quantity of ultraviolet rays would make the test animals less susceptible to rickets, sunlight was filtered through two types of glass boundary: on the one hand ordinary window glass, which was opaque to the so-called Dorno rays between 300 and 313 μm; on the other Vitaglass, which was transparent to these UVB rays, as we would call them today.

The historical juxtaposition of "ordinary" window glass and Vitaglass challenges the narrow, analytical focus on light transparency that still dominates glass studies in cultural history today. [6] "Modernity has been haunted, as we know very well, by a myth of transparency." [7] And this modern myth has literally materialized in glass—in architecture as well as in the sciences. The interwar history of Vitaglass, staged between the window and the test tube, so to speak, subverts the fetish of glass transparency. In straddling the histories of two disciplines,

3 This discursive bump can be visualized by the Google Ngram viewer: https://books.google.com/ngrams/graph?content=Vitaglass&year_start=1900&year_end=1960&corpus=15&smoothing=3&share= (accessed December 23, 2015).

4 John Sadar, "The Healthful Ambience of Vitaglass: Glass, Light, and the Curative Environment," *Architectural Research Quarterly*, 12, no. 3–4 (2008), pp. 269–81. Cf. Daniel Freund, *American Sunshine: Diseases of Darkness and the Quest for Natural Light* (Chicago, IL: University of Chicago Press, 2012), p. 61.

5 Cf. my paper "Environmentality—Revisiting the Milieu of Jakob von Uexküll's *Umweltlehre*," presented at the Annual Conference of the German Studies Association in Denver in October 2013. Cf. Thomas Brandstetter, Karin Harrasser, and Günther Friesinger (eds.), *Ambiente: Das Leben und seine Räume* (Vienna: Turia & Kant, 2010).

6 Kijan Espahangizi, "From Topos to Oikos: The Standardization of Glass Containers as Epistemic Boundaries in Modern Laboratory Research (1850–1900)," *Science in Context*, 28, no. 3 (2015), pp. 397–425. For a differentiated account on glass transparency in the history of modern architecture, see Laurent Stalder, "Glas 1930–1970," in Susanne Hauser and Julia Weber (eds.), *Architektur in transdisziplinärer Perspektive: Von Philosophie bis Tanz—Aktuelle Zugänge und Positionen* (Bielefeld: transcript, 2015), pp. 19–41.

7 Anthony Vidler, "Transparency," in Vidler, *The Architectural Uncanny: Essays in the Modern Unhomely* (Cambridge, MA: MIT Press, 1992), pp. 217–25; here p. 217.

architecture and the experimental sciences, it allows us to gain a deeper understanding of a material boundary technology that has permeated modern society. 8

8 This study therefore meets the methodological requirement to reconstruct a "definite and demonstrable connection, susceptible to theoretical analysis" between the two "cultural environments" of architecture and science—as requested in "Introduction," in Antoine Picon and Alessandra Ponte (eds.), *Architecture and the Sciences: Exchanging Metaphors* (Princeton, NJ: Princeton Architectural Press, 2003), pp. 13–25. Cf. also Peter Galison and Emily Thompson (eds.), *The Architecture of Science* (Cambridge, MA: MIT Press, 1999).

From transparency to the boundary function of glass

The history of modern laboratory research shows that the glittering iconography of the transparent and quasi-immaterial container has obscured from view the rich material historicity and functionality of glass boundary technologies within and beyond the walls of modern laboratories. With the emergence of scientific precision measurement in the nineteenth century and advances in glass science and technology, especially since the 1880s, glass containers turned from handcrafted objects of a usually unknown physicochemical constitution into scrutinized and standardized boundary devices. 9 They were designed to enclose and stabilize artificial environments for specific techno-scientific objects. Chemistry glassware, for example, had to be neutral in regard to the chemical processes contained within it, in order to prevent measurement errors. f.2/f.3 This new physico-chemical knowledge of glass surfaces and boundaries was relevant also to receptacles used in the distribution and circulation of techno-scientific objects in society, such as pharmaceutical ampoules and food containers. 10

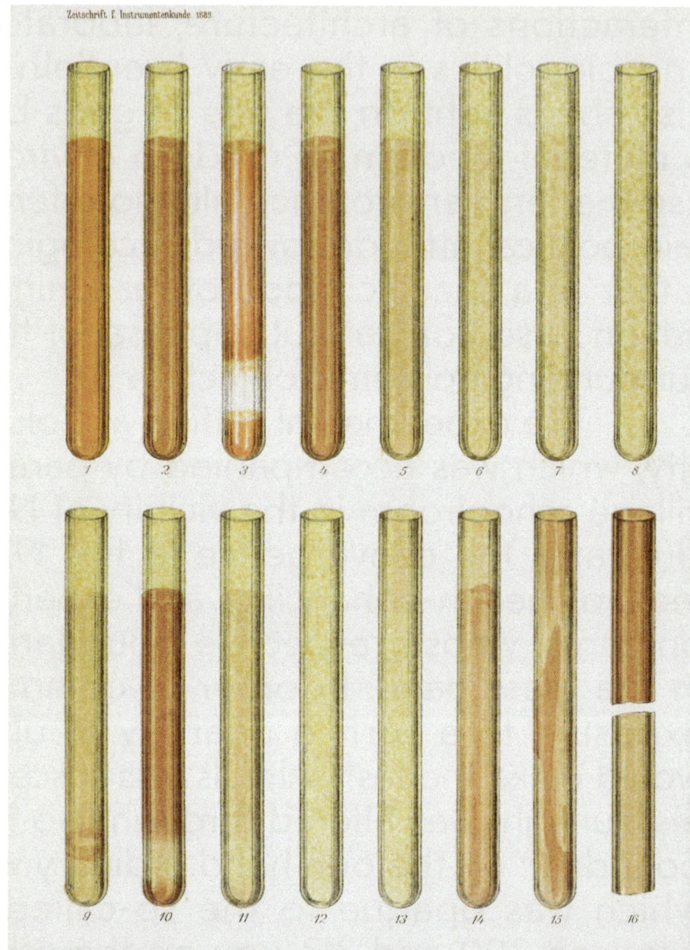

f.2 The chemical resistance of test tubes. A colorimetric test by Franz Mylius, 1889.

9 Cf. Kijan Espahangizi, "From Topos to Oikos" (see note 6), and Espahangizi, "The Twofold History of Laboratory Glassware," in Mathias Grote and Max Stadler (eds.), *Membranes, Surfaces and Boundaries: Interstices in the History of Science, Technology and Culture* (Berlin: Max Planck Institut für Wissenschaftsgeschichte, 2011), pp. 27–44.

10 Kijan Espahangizi, "Immutable Mobiles im Glas: Grenzbetrachtungen zur Zirkulationsgeschichte nicht-inskribierter Objekte," *Nach Feierabend: Zürcher Jahrbuch für Wissensgeschichte*, 7 (2011), pp. 105–28. On new approaches in the historical ontology and epistemology of materials cf. Kijan Espahangizi and Barbara Orland (eds.), *Stoffe in Bewegung: Beiträge zu einer Wissensgeschichte der materiellen Welt* (Zürich/Berlin: diaphanes, 2014).

The leading glass scientist of the American Glass Container Association was well aware of the broad variety of applications developed over the previous decades for new special types of glass. In 1930, he wrote:

"Glass in any form is just glass to many of its users. ... But in reality there exists a wide variety of glasses ... . Adjustments of its chemical composition can change physical properties. Such changes are made to adapt glass for specific uses in industry. ... We can now even obtain special window glass which transmits the

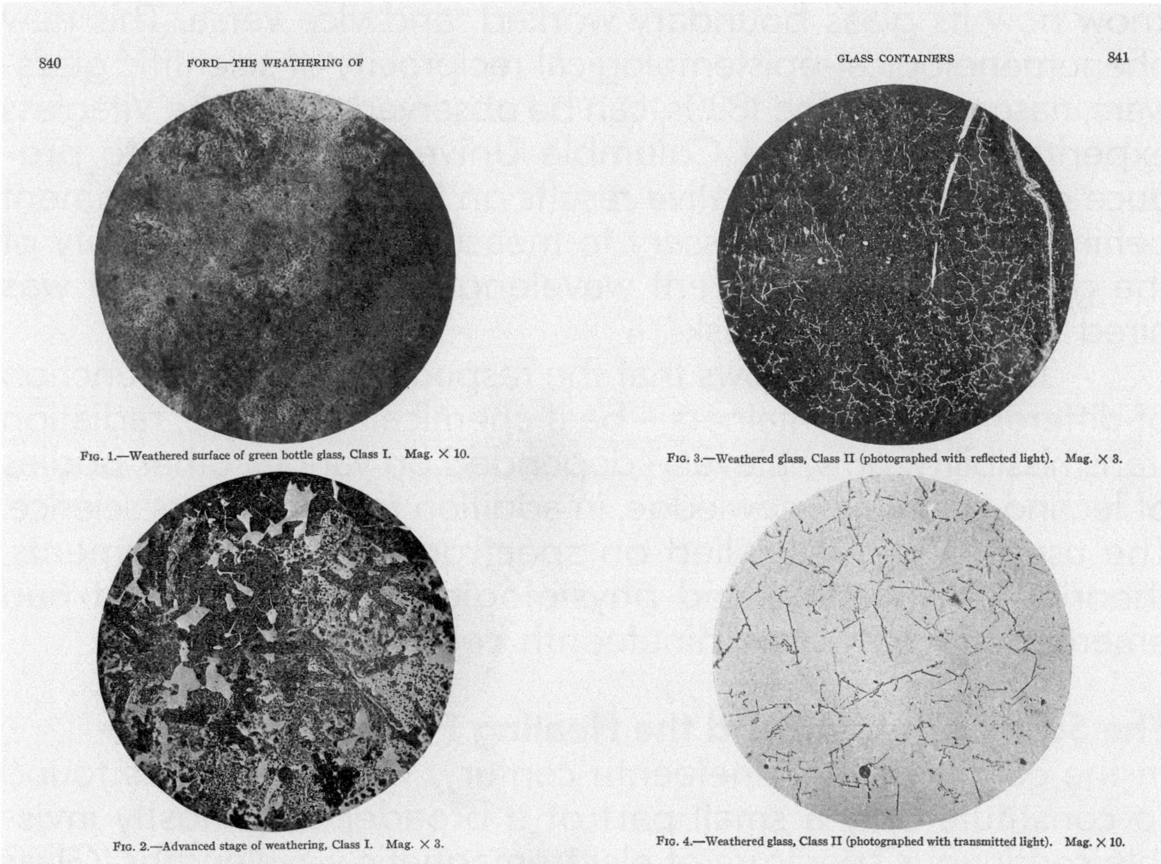
840 FORD—THE WEATHERING OF GLASS CONTAINERS 841

FIG. 1.—Weathered surface of green bottle glass, Class I. Mag. × 10.

FIG. 2.—Advanced stage of weathering, Class I. Mag. × 3.

FIG. 3.—Weathered glass, Class II (photographed with reflected light). Mag. × 3.

FIG. 4.—Weathered glass, Class II (photographed with transmitted light). Mag. × 10.

f.3 Micrographs of weathered glass surfaces, Glass Container Association 1922.

ultra-violet rays of sunlight. Each of these, while of the glass family, shows different properties and widely varying applications. ... Chemical glassware must withstand corrosive action and sudden temperature shock ... . A food container is likewise a special glass. In this article are combined durability, resistance to temperature shock, optical characteristics, color, and strength. ... This glass is the result of extensive studies in the art of glass manufacturing. ... Composition no longer is a matter of tradition but of exact, technically established batch proportions composed of materials selected to give special physical properties. ... Food products can now be held in storage for years in glass without evidence of any effects in the ingredients." [11]

The use of glass containers ranged from scientific research to the packaging of industrially manufactured goods. In the interwar period, this new knowledge of glass entered the architectural field, as in the case of Vitaglass. The types of glass used in these different contexts had in common not only their material substance but also their function. [12] Windows, flasks, test tubes, petri dishes, ampoules, food containers, and milk bottles all had to perform a specific boundary function: to regulate the material, organic, and energetic exchange between an enclosed environment and its surroundings. The new technological boundary function of glass involved the phenomenal as well as the epistemic dimension, which is to say, that in order to know *what* happened within a glass-enclosed environment one had to

11 Karl Ford, "The Rise of the Glass Container," *The Glass Container*, 9, no. 1 (1930), compiled from pp. 5–7.

12 Cf. the chapters "Material" and "Funktion" in Laurent Stalder, "Glas 1930–1970" (see note 6), pp. 25–32; Stalder, "Prä_Liminarien," in Laurent Stalder et al. (eds.), *Schwellenatlas*, *ARCH+*, 191/192 (2009), pp. 24–5.

know *how* its glass boundary worked, and vice versa. This new phenomenological-epistemological reciprocity of scientific glassware, nascent since the 1880s, can be observed also in the Vitaglass experiments of 1928 at Columbia University. In order to produce meaningful quantitative results on the healthy environment behind glass it was necessary to measure the transmissibility of the glass panes at different wavelengths. A MIT physicist was hired to carry out this task. [13]

13 Eddy, "Ultra-Violet Light Transmitting Windows" (see note 1), p. 1474.

This example shows that the respective boundary function of different glass containers — be it chemical resistance, radiation transmissibility, or whatever — depended on various other bodies of techno-scientific knowledge, in addition to actual glass science. The use of Vitaglass relied on spectrographic measurements, theories of radiation, and physiological knowledge that had emerged since the late nineteenth century.

The Science of Light and the Healing Power of UV Rays

In the course of the nineteenth century, visible light was found to constitute only a small part of a broader and mostly invisible continuous spectrum of electromagnetic wavelengths. Glass devices such as prisms, lenses, and glass screens played a crucial role in this development. As early as the 1820s, scientists were aware that glass, while transparent to visible light, was hardly transmissible to invisible heat and ultraviolet rays — or, as the latter were commonly known at the time, chemical or actinic rays. [14] Glass screens were used as selectively transmissible filters in order to study the effects of different parts of the spectrum, for example in bacteriology.

14 Cf. Klaus Hentschel, "Macedonio Melloni über strahlende Wärme," *NTM*, 13 (2005), pp. 216–37.

In 1877, two British scientists placed test tubes with bacteria solutions "on the window-ledge" of their laboratory and observed that this exposure to sunlight destroyed the microbes. [15] They used colored glass filters in order to find out which part of the spectrum of sunlight was responsible for this effect. It took until the end of the nineteenth century to show that the germicidal effect was due to ultraviolet radiation. Other "healthy" effects of sunlight and especially of ultraviolet radiation were discovered too, at that time, such as enhanced tissue growth, and the healing of eczema, lupus, and other diseases. [16] The invention of a lamp for artificial UV rays facilitated research into their effects in chemistry, physics, biology, and medicine, most importantly in dermatology and heliotherapy. One drawback to these lamps was that they were based on the electrical discharge of mercury vapor in an ordinary soda-lime glass envelope that absorbed a great deal of the UV rays. In 1904, Otto Schott, who was the most influential glass scientist at the time, and owner of the

15 Philipp Hockberger, "The Discovery of the Damaging Effect of Sunlight on Bacteria," *Photochemistry and Photobiology, B*, 58 (2000), pp. 185–91; Hockberger, "A History of Ultraviolet Photobiology for Humans, Animals and Microorganisms," *Photochemistry and Photobiology*, 76, no. 6 (2002), pp. 561–79.

16 For the detailed history of light and UV therapy, see Niklaus Ingold, *Lichtduschen: Geschichte einer Gesundheitstechnik 1890–1975* (Zurich: Chronos, 2015).

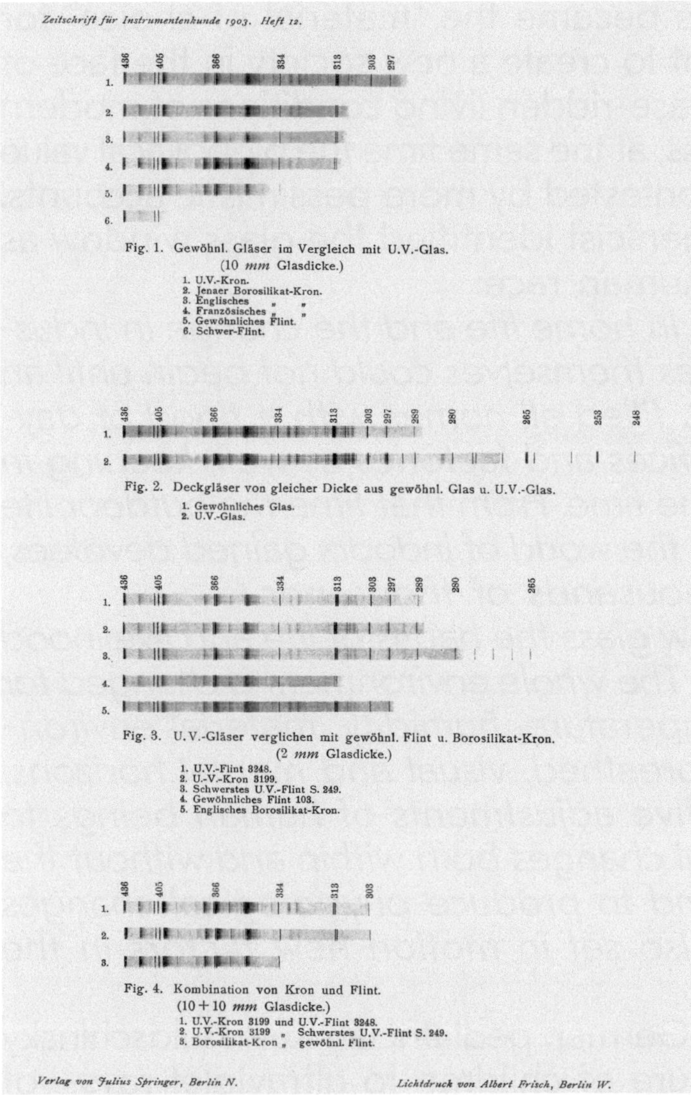

f.4 Comparison of ordinary glass and UV glass by Eberhard Zschimmer, 1903.

world's leading company for scientific glassware, presented a new glass composition that would allow the healthy rays to leave the lamp. Schott was able to draw on the broad expertise in glass technology and scientific glassware garnered by his company since the 1870s. This "Uviolglas," as it was known, had been developed in 1903 by one of Schott's most proficient glass scientists, Eberhard Zschimmer, mainly for optical uses. 17 Studies since the mid-nineteenth century had shown that optical lenses absorbed the invisible UV rays in the light spectrum. This posed a serious problem for all experimental sciences using spectrographic methods, such as chemistry and astronomy. 18 In other words, progress in the science of light depended on optical media that were transmissible both to visible and invisible rays. f.4 Zschimmer's new glass, in the form of container glass and windows, was also used for physiology and light therapy. So, evidently, when Vitaglass arrived on the scene in the mid-1920s, it wasn't the first glass product transmissible to UV rays. How, then, did it come to be a major public issue in the interwar period?

17 Eberhard Zschimmer, "Über neue Glasarten von gesteigerter Ultraviolett-Durchlässigkeit," *Zeitschrift für Instrumentenkunde*, 23, no. 12 (1903), pp. 360–3.

18 Cf. Otto Schott, "Über eine neue Ultraviolett-Quecksilberlampe (Uviollampe)," *Zeitschrift für angewandte Chemie*, 18, no. 16 (1905), pp. 615–22. Uviol was a commercial product, so Schott did not reveal the composition of the glass.

Public health, glass architecture, and Vitaglass

After the Great War, a broad public health and urban sanitation movement developed on both sides of the Atlantic Ocean. The main preoccupation of the advocates of this so-called "social hygiene" was the "influence of natural and artificially produced climatic conditions," i.e. of air, water, temperature, and especially sunlight, on the population. 19 The new avant-garde of glass architecture that had gathered shortly before the First World War shared (and fueled) this excessive contemporary passion for sunlight. 20

19 A[dolf] Gottstein, A[rthur] Schlossmann, and L[udwig] Teleky (eds.), *Handbuch der sozialen Hygiene und Gesundheitsfürsorge*, vol. 5: *Soziale Physiologie und Pathologie* (Berlin: Springer, 1927), pp. 146, 125.

20 Cf. Paul Scheerbart, *Glasarchitektur* (Berlin: Der Sturm, 1914), chap. 1, and again Sadar, "Healthful Ambience of Vitaglass" (see note 4), pp. 269–72.

In the interwar years, glass became the "material of choice" for those architects who sought to create a new society in the face of the dark, polluted, and disease-ridden living conditions of modern urban spaces. 21 Nonetheless, at the same time the biopolitical value of glass architecture was contested by more pessimistic accounts. In 1919, a US-American eugenicist identified the glass window as a veritable threat to the human race:

"[T]he great change in home life and the change in industrial life and in the industries themselves could not begin until an abundance of cheap glass filled all homes with a flood of daylight, and all shops and offices and factories as well, keeping in the artificial heat at the same time. From that time the outdoor life rapidly lost its people while the world of indoors gained devotees, willing or unwilling, by thousands of thousands."

Further: *"With window glass the habits of life and livelihood are completely changed. ... The whole environment is changed for the species, including temperature, humidity, material environment, composition of air breathed, visual and mental horizons, and a change in the relative adjustments of human beings to disease germs. Such radical changes both within and without the human organism are bound to produce physiological changes in the individuals. They also set in motion new factors in the evolution of the race."* 22

In 1918, however, the German pediatrician Kurt Huldschinsky discovered that the exposure of children to ultraviolet rays—of the natural (sunlight) and artificial (mercury arc lamp) variety—could cure rickets. Especially given the deprivations of the First World War, this disease, which turned out to be a Vitamin D deficiency 23 that caused severe bone deformities, stunted growth, and infant mortality, had become pandemic among children, especially in cities. Huldschinsky's groundbreaking discovery of the antirachitic power of UV rays, however, seemed to deliver a strong argument against such an alarming interpretation of glass architecture. One can only imagine the shock when the public realized that the utopia of new glass architecture had a major flaw: since ordinary window glass was virtually opaque to ultraviolet rays, the most vital part of the sunlight spectrum never reached the living room. No wonder *Time* magazine was appalled: "Children and animals that live in glassed houses are cheated of that ultraviolet part of the sun's light which helps the bones ossify. The glass blocks the ultraviolet rays; the children and animals become rickety." 24 A German newspaper warned its readers: "Our windows are harmful to health!" 25 As mentioned above, scientists had known since the first half of the nineteenth century that ordinary soda-lime glass was opaque to UV rays, but

21 Sadar, ibid, p. 270; Stalder, "Glas 1930–1970" (see note 6).

22 R. E. Danforth, "Window Glass as a Factor in Human Evolution," *The Scientific Monthly*, 8, no. 6 (1919), pp. 537–557; here pp. 537, 540.

23 Kurt Huldschinsky, "Heilung von Rachitis durch künstliche Höhensonne," *Deutsche Medizinische Wochenschrift*, 45 (1919), pp. 712–3. Hockberger, "History of Ultraviolet Photobiology" (see note 15), p. 565.

24 "Sun & Glass" (see note 2).

25 "Unsere Fensterscheiben sind gesundheitsschädlich, denn sie lassen die segensreichen Ultraviolettstrahlen nicht durch! Die Wissenschaft auf der Suche nach einem gesundheitsförderlichen Fensterglas," *Abendblatt der National-Zeitung*, December 15, 1927.

it was only now that their expert knowledge pervaded the public debate on account of its biopolitical relevance. In no time at all, a range of new remedial glass products "appeared for sale," most prominently Vitaglass. 26

Vitaglass was invented by Francis Everard Lamplough, a glass technologist from England, and registered as a trademark in 1925. The low level of iron oxide in the Vitaglass composition increased its UV transmissibility. 27 At the initiative of physiologist Leonard Hill of London University, an expert for the effects of climate on health, Vitaglass was first tested at the monkey house of the city zoo. Several glass companies in the UK and the US immediately embarked on the mass production of Vitaglass, as a joint venture. To promote international sales, they set up a Vita Glass Marketing Board and developed a "dedicated" propaganda strategy. 28 It was their advertising campaign that had initiated the public debate on the UV transmissibility of window glass and public health in 1926 in the first place. But the media focus on Vitaglass unintentionally prompted the launch of rival products on this new market, among them Helioglas, Sanalux, Ultravit, and Corex, and many others. 29 When the German glassworks in Jena found out about the new Anglo-American Vitaglass fad its response was immediate: it developed three new Uviolglas products, started mass production, lowered its prices, and initiated a marketing campaign of its own. Uviol advertisements appeared in no less than twenty-five professional journals for medicine, architecture, gardening, horticulture, or animal farming, as well as in several German-language newspapers. 30 Not surprisingly, the ads emphasized that Uviolglas had been invented "25 years ago" and used occasionally in sanatoriums and gardening prior to the First World War. On the one hand, Schott berated the "blatant advertising abroad" for ignoring the German invention of 1903; on the other, his company recognized the benefits of the "immoderate" Anglo-American media hype: it put UV glass back in the public eye. 31

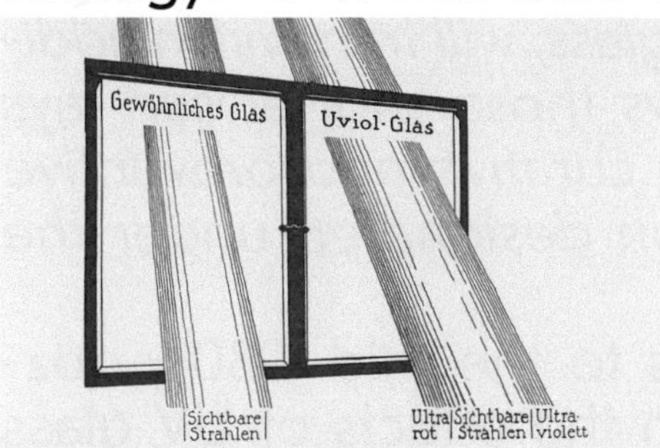

f.5 Uviol advertisement, Schott 1928.

Publicity for an appealing and commercially successful UV glass was one thing, but manufacturing it was another. The various products competed in regard to their transmission rate (and its consistence over time), and to their price; and they also had to meet architectural requirements, such as mechanical stability, suitability for mass production, resistance to weathering, and so forth. In spite of these practical challenges, the idea behind the new products fell on fertile soil both among the general public and the scientific community. f.5

26 "Sun & Glass" (see note 2).

27 Between 1908 and 1914, William Crookes had found out that glasses containing a high level of iron oxide blocked UV rays: "The Preparation of Eye-Preserving Glass for Spectacles," *Philosophical Transactions of the Royal Society of London*, A 214 (1914), pp. 1–25.

28 On the history of Vitaglass Corporation and their marketing strategy, cf. Sadar, "Healthful Ambience of Vitaglass" (see note 4), pp. 272–81.

29 "UV-Berichte IV," Schott Archive Jena 7/62 (Laborberichte 35, 1917–1944), p. 7. Cf. H[arrison] P. Hood, "A New Ultra-Violet Transmitting Glass," *Science*, 64 (New Series), no. 1655 (1926), pp. 281–2.

30 "Das Ultraviolettglas," *Der Tag*, February 9, 1928; "Fensterglas und Sonne," *Münchener Neueste Nachrichten*, February 13, 1928; "Glas als Arzt: Therapeutisches Fensterglas," *Deutsche Allgemeine Zeitung*, May 20, 1928; "Das Jenaer Uviolglas," *Wiener Handelsblatt*, September 4, 1928.

31 "Das ultraviolettdurchlässige Fensterglas ist eine 25 Jahre alte deutsche Erfindung," "Jenaer Uviol-Glas 4262," "Über Sonnenstrahlen und Ultraviolett-Durchlässiges Glas—Mitteilung 4270," Schott Archive Jena 18/12 (Das Jenaer Uviolglas 1908–1930).

The scientific value of UV glass

The ambitious commercial advertising induced research that aimed to thoroughly evaluate the possible benefits of UV glass products. Scientists such as Walter H. Eddy, who conducted the Vitaglass experiments at Columbia University, were conscious of this motivation:

"A NEW industry and a new line of selling talk have sprung into being within the past few years. ... The selling talks of salesmen endeavoring to convince prospective customers demand critical evaluation and constructive suggestion to eliminate claims made without basis in fact.

The industry to which I refer is that concerned in producing window material which, correcting a defect of ordinary window glass, will transmit in addition to the illuminating rays those shorter sun rays that are now known to be curative and preventive of the faulty lime utilization designated under the term 'rickets'." [32]

From the mid-1920s to the mid-1930s, dozens of scientific studies on the effects of UV glass were carried out on both sides of the Atlantic, in clinical studies, bacteriology, zoology, agriculture, and so forth. They included tests with rachitic rats behind Vitaglass, studies on the height, weight, hemoglobin level, and bone growth of the children in vita-glazed schoolrooms, [33] studies on crops grown in vita-glazed greenhouses, [34] studies that measured the concentration of bacteria in the air of vita-glazed rooms, [35] and many more. [36] In the same period, studies were made of how light and UV radiation might modify foodstuffs, beverages, and pharmaceutical substances enclosed in glass containers, thereby referencing the general functionality of glass as a boundary material for all kinds of techno-scientific objects, from chemical substances to the modern living room. [37]

Even before these scientific studies were able to produce well-founded knowledge, clinics, zoos, greenhouses, schools, and sanatoriums on both sides of the Atlantic started replacing their windows. [38] Evidently, public faith in the benefits of UV glass windows was widespread and strong. Schott's glassworks even used its customer base to gather

32 Eddy, "Ultra-Violet Light Transmitting Windows" (see note 1), p. 1470.

33 Ian Sutherland, "Growth under Vita Glass," *Journal of Hygiene*, 32, no. 2 (1932), pp. 211–18.

34 W. Dix, "Vegetationsversuche mit ultraviolette Strahlen durchlässigem Glas," *Gartenbauwissenschaft*, 2 (1929), pp. 365–8; J. Reinhold and M. Schmidt, "Versuche mit verschiedenen Glassorten im Frühgemüsebau," *Gartenbauwissenschaft*, 3 (1930), pp. 301–30. See the research overview in H. W. Popp and Florence Brown, "A Review of Recent Work on the Effect of Ultraviolet Radiation Upon Seed Plants," *Bulletin of the Torrey Botanical Club*, 60, no. 3 (1933), pp. 161–210.

35 Jean Broadhurst and Theodore W. Hausmann, "Bacterial Destruction Through Glass," *The American Journal of Nursing*, 30, no. 11 (1930), pp. 1391–4.

36 Cf. "Literatur über biologische Wirkung von UV-Strahlen usw.," Schott Archive Jena 18/12 (Das Jenaer Uviolglas 1908–1930).

37 In 1926, a Glass Container Committee was appointed by the American Pharmaceutical Association in order to investigate the "deterioration of certain chemicals by light rays," especially ultraviolet rays. Cf. also Gustav Keppeler, "Untersuchungen an Flaschengläsern," *Glastechnische Berichte*, 8, no. 2 (1930), pp. 65–77; Deichmüller, "Die schädlichen Wirkungen ultravioletter Strahlen und die Lichtdurchlässigkeit, insbesondere bei transparenten Folien," *Die Verpackung*, 7, no. 18 (1932), pp. 283–4.

38 Cf. Sadar, "Healthful Ambience of Vitaglass" (see note 4), and Kijan Espahangizi, *Wissenschaft im Glas: Eine historische Ökologie moderner Laborforschung* (diss. ETH Zurich, 2010), p. 305, fn. 768.

more information, requesting nurseries, schools, agricultural businesses, and factories from all over the country to report on their experience of UV glass. [39] But the feedback was ambiguous, as in most scientific studies on UV glass. This was also the case with the tests on rachitic rats at Columbia University, even though there were in fact noticeable effects: the bones of the test animals behind Vitaglass differed from those behind ordinary glass. "Camera lucida drawings of the split ends of 'rat knee bones" illustrated this effect. [40/f.6]

The general problem with most UV glass studies and the reason for their ambivalence in regard to possible benefits was that experimental results were attained in a highly controlled environment and could therefore not simply be transferred to real-life situations, where the relevant parameters—such as the percentage of skin surface covered by clothes, the distance to the window, the paths of rays, the ratio of reflected rays, etc.—were far more complex and variable. [41] All these aspects affected the assessment of the cost-benefit-ratio rather negatively. The main argument against the actual installation of UV glass windows especially in urban areas, however, was that, before the sunlight reached the windows, a substantial number of the UVB rays would be filtered out by the atmosphere, climatic conditions, or smog, [42] to say nothing of the dirt that usually clouded the mundane reality of glass architecture, as Lewis Mumford observed in the late 1930s. [43]

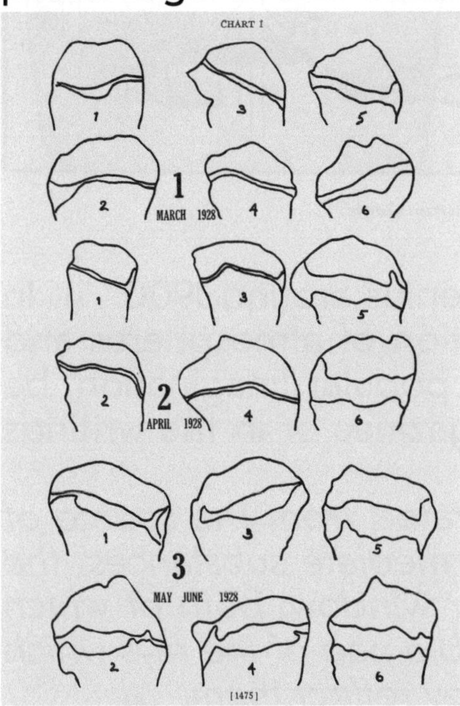

Vitaglass and environmentality

Despite its suggestive power and the general buzz it created, Vitaglass was a commercial failure. Yet, if considered from a broader historical and epistemological perspective, the interwar debate on the UV transmissibility of glass windows was astonishingly productive. It served to interrelate and restructure current knowledge of different environmental boundaries: the natural ones such as the atmosphere and the technical ones made of glass.

The histories of glass manufacture and of the atmospheric sciences have been deeply intertwined since the early nineteenth century. The idea that the earth's atmosphere functions like a giant glasshouse that is transparent to visible light but opaque to heat rays was first hypothesized by Jean Baptiste Fourier in the

39 Cf. Schott Archive Jena 18/12 (Das Jenaer Uviolglas 1908–1930).

40 Walter Eddy, "Ultra-Violet Light Transmitting Windows" (see note 1), p. 1474: "The narrowness of the 'line' between the end of the bone and the shaft indicates degree of protection. The narrower the line the higher the protection. ... The numbers 1, 2, 3, 4 indicate animals that were in front of Vitaglass. The numbers 5 and 6 indicate animals that were in front of ordinary window glass."

41 Ibid., p. 1472.

f.6 The knees of rachitic rats behind Vitaglass, 1928.

42 Janet H. Clark, "The Probable Amount of Ultra-Violet Radiation Obtained Indoors Through Ultra-Violet Transmitting Glass," *Science*, 68 (New Series), no. 1755 (1928), pp. 165–6; Konrad Büttner and Erika Sutter, "Der Einfluss des Großstadtdunstes auf das Strahlungsklima, insbesondere im Ultraviolett', *Die Naturwissenschaften*, 17, no. 33 (1929), p. 652.

43 Lewis Mumford, *The Culture of Cities* (New York: Twayne, 1998 [1938]), pp. 207–8. For an early indication see "Ultra-Violet Light Transmitting Glass and City Smoke," *Glass Review*, 3, no. 2 (1927), p. 23.

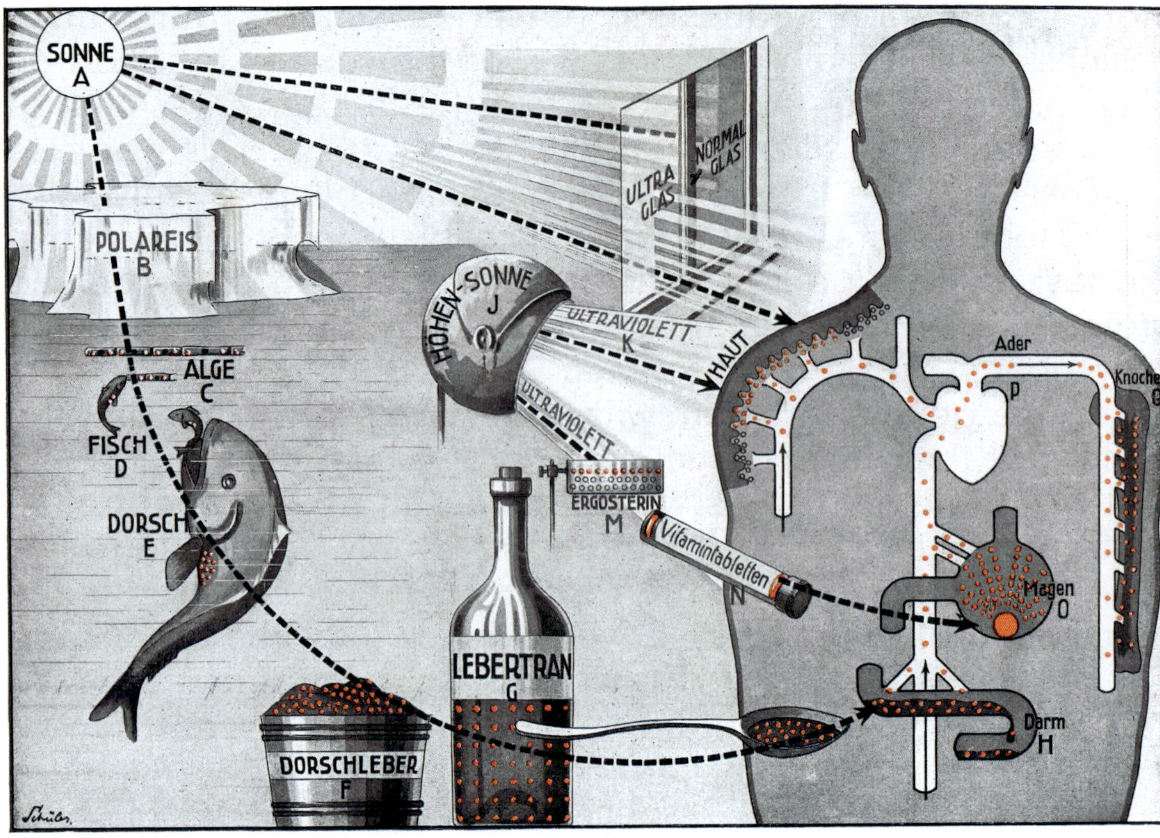

Die biologischen Wirkungen des ultravioletten Lichtes.
Erklärung im Text S. 254/255.

f.7 Fritz Kahn, "The Biological Effects of Ultraviolet Light," 1926.

1820s then popularized by Svante Arrhenius around 1900. 44 In the 1920s and 1930s, the close association of atmospheres and glass envelopes became a part of the popular imagination, be it on the covers of American pulp magazines or in the writings of glass architects:

"Our enclosed spaces are separated from the source of all light on earth, the sun, by two intermediate substances, the atmosphere and the glass layer of our window, both of which are not passed without loss. Both absorb a part of the rays which originate from the sun or the sky, or they reflect them." 45

It is safe to say that glass is an important historical medium in which concepts of environmental boundaries both fictitious and real have taken form since the nineteenth century—literally materialized in glass. 46 In this sense, the UV glass debate of the interwar period may be read as a productive episode in the material genealogy of modern environmentality.

First of all, it involved different actors and issues—the experimental sciences, architecture, public health, biopolitics, industrial production, and so on—in establishing a new socio-ecological infrastructure. 47 Secondly, it changed the very functionality of the environmental glass boundary. Fourier's glasshouse had been a naturally given semipermeable filter. One could use it, but not change the parameters. The interwar debate on UV radiation—inspired by earlier developments in experimental

44 Espahangizi, "The Twofold History of Laboratory Glassware" (see note 9), p. 37; Svante Arrhenius and H. Borns, *Worlds in the Making: The Evolution of the Universe* (New York: Harper, 1908), p. 52.

45 Otto Völkers, *Glas und Fenster: Ihr Wesen, ihre Geschichte und ihre Bedeutung in der Gegenwart* (Berlin: Bauwelt Verlag, 1939), p. 76. See, for example, the urban glass dome on the cover of *Wonderstories* (Gernsback: New York, 1935).

46 Cf. the general argument in Espahangizi, "Twofold History of Laboratory Glassware" (see note 9). In this sense, the object of glass supports the argument that the spaces in which scientific practice takes place not only shape the conception of the sciences but also interact with the practice of architecture, developed in Galison and Thompson (eds.), *Architecture of Science* (see note 8).

47 Cf. Paul N. Edwards, "Infrastructure and Modernity: Force, Time, and Social Organization in the History of Sociotechnical Systems," in Thomas J. Misa, Philip Brey, and Andrew Feeberg (eds.), *Modernity and Technology* (Cambridge, MA: MIT Press, 2004), pp. 185–225; here p. 188–9.

research since the 1880s—fostered the perception of glass as an adjustable boundary device.

Therefore, the Vitaglass debate must be understood as an important episode in the prehistory of the environmental turn in twentieth-century architecture described by Reyner Banham. Revealingly, it took place at some remove from the avant-garde discourse of modern glass architects. 48 UV glass windows performed the "operative transparency" of today's hi-tech glass façades *avant la lettre*—which is to say, they regulated system-environment relations rather than giving the building a "face." 49

A third productive outcome of the UV glass debate of the 1920s was its contribution to the broader undercurrent of modern environmentality beyond the architectural field. This consisted in its integration of various epistemic, technical, and metaphorical notions of environmental boundaries in one holistic image: the atmosphere, the human skin, containers for objects, or architecture. Fritz Kahn's epic work of the mid-1920s on the modern human body offers a very telling example of this new socio-ecological imagination in the early twentieth century. f.7 In the upper part of the illustration we see the UV glass iconography. Both the windowpane and the glass container are incorporated as boundary devices in the metabolic pathways of the health-enhancing Vitamin D. In this picture we can grasp the emerging notion of an overarching, multi-layered environmental "living machine" that is structured by natural as well as man-made technological boundaries and in which modern human existence is embedded. 50 This liminality—epitomized by the glass boundary between organicist and mechanistic interpretations, between an environmental "diode" and a "second skin"—runs like a thread through the architectural debates of the twentieth century. 51 Against this backdrop, the UV glass debate of the interwar period reveals itself to be a productive historical conjuncture in the material, epistemic, and imaginative genealogy of modern environmentality.

Post scriptum: The Biosphere 2 project—designed as a simulation of the real biosphere and built in the desert in Arizona in 1991—"rediscovered" the poor UV transmissibility of ordinary glass. The lack of UV radiation negatively affected the vegetation in the artificial ecosystem. 52

48 Reyner Banham, *The Architecture of the Well-Tempered Environment* (London: Architectural Press, 1984 [1969]), p. 18.

49 Nikolaus Kuhnert and Angelika Schnell, "Transparencies Yet to Come," *ARCH+*, 144/145 (1998), pp. 18–19; here p. 19. Cf. Völkers, *Glas und Fenster* (see note 45), pp. 73–80.

50 Cf. Völkers, *Glas und Fenster* (see note 45), p. 78; Ted Krueger, "Like a Second Skin: Living Machines," in *Architectural Design*, 66, no. 9/10 (1996), pp. 29–32.

51 Krueger, ibid, p. 29; Mike Davies, "A Wall for all Seasons," *Royal Institute of British Architects*, 88, no. 2 (1981), pp. 55–7.

52 Charles Cockell, Adrian Southern, and Aleshs Herrera, "Lack of UV Radiation in Biosphere 2: Practical and Theoretical Effects on Plants," *Ecological Engineering*, 16 (2000), pp. 293–9.

Sound Conduits: Displaying the Architecture of Telecommunications

Carlotta Darò

Carlotta Darò is Maître-assistant at the École nationale supérieure d'architecture Paris-Malaquais and at the Laboratoire LIAT.

The inventions of the telegraph, the telephone, radio—even the phonograph—are usually associated with ingenious figures and powerful tales that emphasize how those machines suddenly changed both our perception of sound and the very way in which we hear. "Mr. Watson, come here. I want to see you!" said Alexander Graham Bell to his assistant in 1876, through the first telephone connection. Thomas Edison heard his own voice reciting a popular nursery rhyme—"Mary had a little lamb"—returning to him from the cylinder of the first ever phonograph, built in 1878. "Are you ready?" was, for its part, the first wireless message transmitted in Morse code by Guglielmo Marconi in 1897. That very same year, the inventor and entrepreneur founded his own company. Social and cultural revolutions—we are told—are related, then, to specific turning points that allow sound to lose a little bit of its ephemeral character and take people into the past or across long distances. As Jonathan Sterne points out, these are powerful stories, because, by fixing a moment and by building a sort of myth, they tell us that something happened to the nature of sound during the late nineteenth century. [1] However, these tales are also historically incomplete since they tend to obliterate the practices, technologies, and ideas that emerged prior to these inventions. In our collective imaginary, sound reproduction technologies, such as the telephone and radio, are closely related to the invention of the ultimate device, the receiver or sound machine that allows a whole electric or electromagnetic system to operate. Yet it is useful to move beyond such devices and take another, broader perspective. The purpose of this study is therefore to focus on the conduits that power telephone and

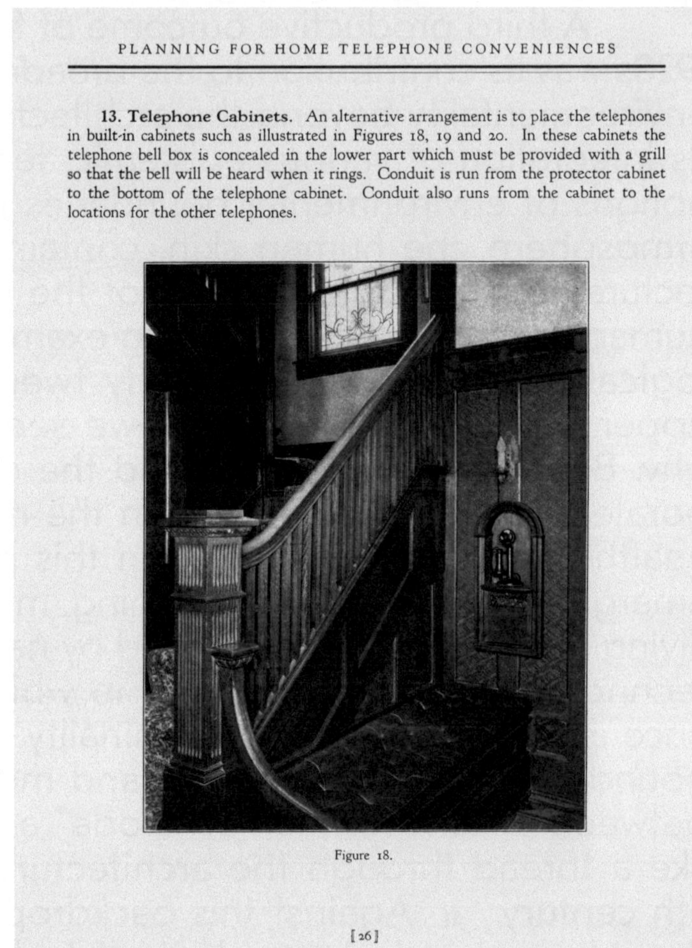
PLANNING FOR HOME TELEPHONE CONVENIENCES

13. Telephone Cabinets. An alternative arrangement is to place the telephones in built-in cabinets such as illustrated in Figures 18, 19 and 20. In these cabinets the telephone bell box is concealed in the lower part which must be provided with a grill so that the bell will be heard when it rings. Conduit is run from the protector cabinet to the bottom of the telephone cabinet. Conduit also runs from the cabinet to the locations for the other telephones.

Figure 18.

[26]

f.1 *Planning for Home Telephone Conveniences*, AT&T, 1928.

1 Jonathan Sterne, *The Audible Past* (Durnham: Duke University Press, 2003), pp. 1–2.

radio systems, as well as on how these became a part of architectural practice through their integration in planning processes. To this end, the material considered remains within a small scale, that is to say, the home or even the apartment. In addition, attention is paid here to original documents produced by American telecommunications companies at the height of their popularity, from the late 1920s to the 1940s, specifically to appeal to architects, engineers, and consumers.

A 1928 American Telephone and Telegraph Company (AT&T) booklet, *Planning for Home Telephone Conveniences*, illustrates the growing popularity of telephone services and their ongoing installation in American homes, which increasingly required the development of innovative design strategies from the range of people—owners, architects, engineers, and builders—involved in the construction process: "The placing of facilities for telephone wires and apparatus during construction results in general improvement of the residence in that better appearance is secured by concealing the wires and some of the apparatus." 2 A new science of planning emerged to address the integration of concealed or visible telephone facilities and apparatuses that were conceived either for immediate or future installation. "Conduits, protector cabinets, telephone cabinets, or other arrangements for concealing the wires and apparatus," the booklet says, "must necessarily become a permanent part of the building." 3 In other words, installing telecommunications services was now not just the business of technicians and engineers but also an architectural concern. The booklet goes on to enumerate technical and functional details through a multitude of examples. Telephone conduits, wires, cables, outlets, and plugs had to enter the domestic interior as discreetly as possible. Rigorously separated from the electrical wiring, telephone conduits usually entered buildings through the exterior walls,

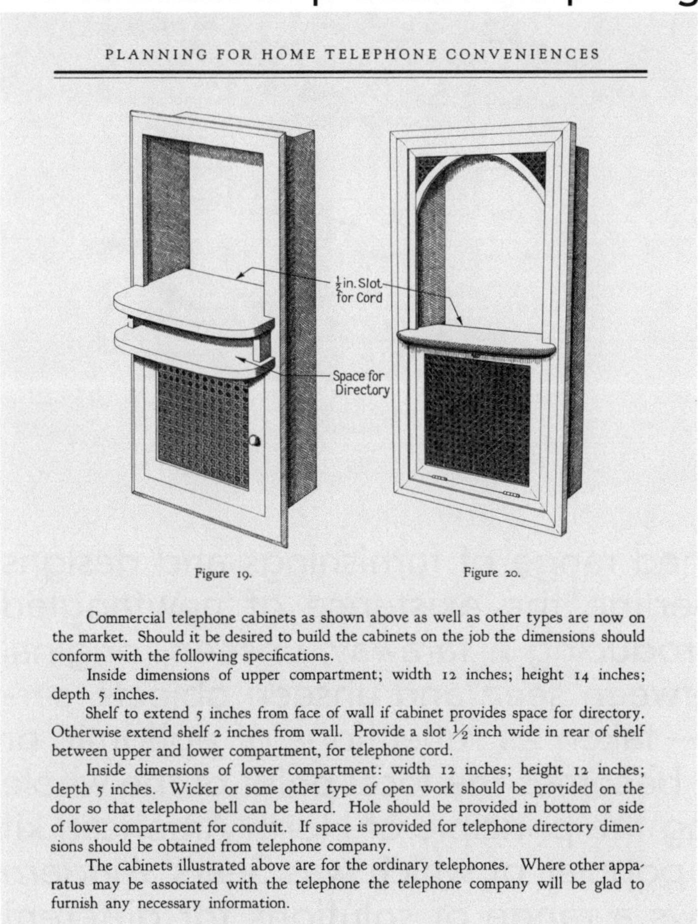
PLANNING FOR HOME TELEPHONE CONVENIENCES

Figure 19. Figure 20.

Commercial telephone cabinets as shown above as well as other types are now on the market. Should it be desired to build the cabinets on the job the dimensions should conform with the following specifications.

Inside dimensions of upper compartment; width 12 inches; height 14 inches; depth 5 inches.

Shelf to extend 5 inches from face of wall if cabinet provides space for directory. Otherwise extend shelf 2 inches from wall. Provide a slot ½ inch wide in rear of shelf between upper and lower compartment, for telephone cord.

Inside dimensions of lower compartment: width 12 inches; height 12 inches; depth 5 inches. Wicker or some other type of open work should be provided on the door so that telephone bell can be heard. Hole should be provided in bottom or side of lower compartment for conduit. If space is provided for telephone directory dimensions should be obtained from telephone company.

The cabinets illustrated above are for the ordinary telephones. Where other apparatus may be associated with the telephone the telephone company will be glad to furnish any necessary information.

[27]

f.2 *Planning for Home Telephone Conveniences*, AT&T, 1928.

2 *Planning for Home Telephone Conveniences: For the Use of Architects, Engineers, Builders, Owners* (American Telegraph and Telephone Company, 1928), p. 7.

3 Ibid.

moving from the ground up, through the cellar ceiling or the first floor of the building. More rarely, they came in from above by passing through the attic. Telephones within domestic interiors were typically displayed on specially designed cabinets, or on already existing tables that comprised a built-in bell box cabinet, or they took the form of a wall-mounted handset telephone; and comfortable sofas and chairs accompanied them always, so as to foster the homely integration of this newly technologized practice of oral communication. f.1/f.2

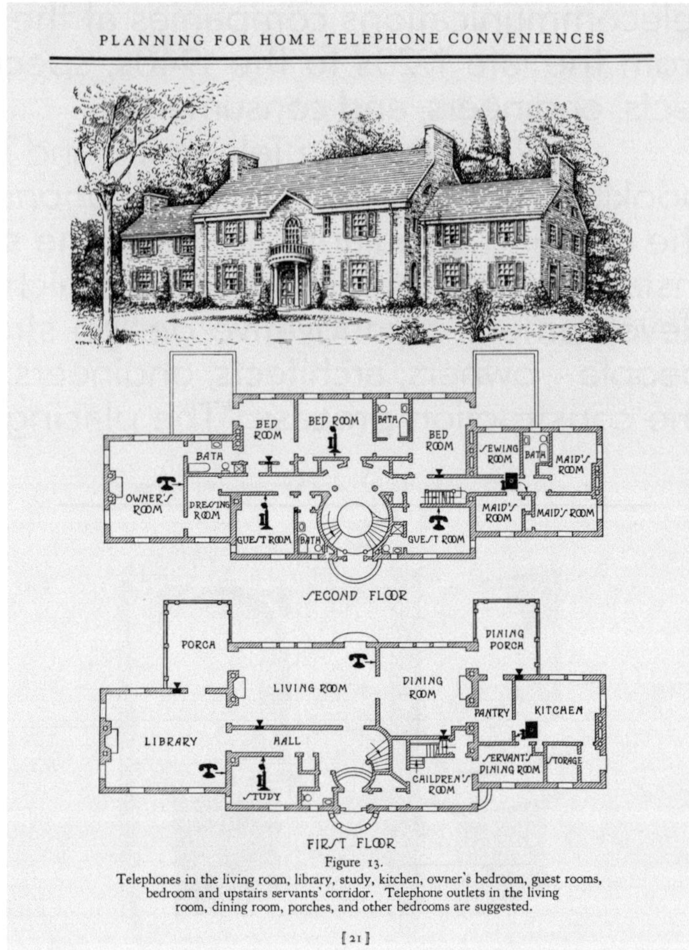

f.3 *Planning for Home Telephone Conveniences*, AT&T, 1928.

The widespread infatuation with sound reproduction technologies resulted in a paradoxical need to display these emblems of modern acousmatic sound ("acousmatic" being the common designation of a sound that one hears without seeing its original source). 4 As the material manifestation of invisible communication, the aforementioned range of furnishings and designs came to frame and underline the existence of newfangled machines capable of reproducing a faraway, unseen, original sound. This distinction between seen and unseen objects, services, and infrastructures—taken as reflections of epitomic or limited functional means—becomes the focal point of the whole AT&T publication. Following the principle of North American kit house catalogs (the most popular of which was Sears' *Modern Homes*), the booklet shows a range of solutions for different homes, family configurations, and needs. From the single-family residence to the duplex or multi-unit apartment building, from the modernist to the vernacular or country-home style, the authors propose a specific installation of telephone facilities in order to obtain the most functional, comfortable, and efficient service. In addition to the visible system, that is to say the variously emphasized speaking and hearing points, the booklet displays plans for the attendant telephone conduit layout, the concealed side of a network f.3/f.4 —for even though hidden away, such conduits

4 On the definition and divulgation of the notion of acousmatic sound, see Pierre Schaeffer, *Traité des objets musicaux: Essai interdisciplines* (Paris: Éditions du Seuil, 1966), and Michel Chion, *Guide des objets sonores: Pierre Schaeffer et la recherche musicale* (Paris: Buchet-Chastel, 1983). For a critical investigation of the subject in terms of practices and culture, see Brian Kane, *Sound Unseen: Acousmatic Sound in Theory and Practice* (Oxford: Oxford University Press, 2014).

now become pivotal to the workings of the entire system and must therefore be accurately planned in advance. Furthermore, while telephones are somehow fixed and standardized objects, conduit layouts are the flexible part of the system; and crucially so, since they have to be adapted to specific needs and spatial configurations.

That same decade, AT&T published a second, less technical, and more captivating version of its guide to installing telephone facilities. This time, the booklet insisted on the explicit association with modernity and modern needs, addressing "the modern American," and focusing on the "modern concept of comfortable living." 5 At the time, sound reproduction technologies represented the promise of science and rationality. They opened up new temporal patterns of communication in everyday life and were considered, in a word, modern. 6 Indeed, visible apparatuses such as telephones, radios, and the furniture that went with them embodied this relation to modernity. They became, as Adrian Forty would later put it, "objects of desire," and their technical appearance an exciting emblem of progress. 7 However, what is especially striking in this case is that, beyond the telephone itself, taken as an autonomous machine, the true object of desire seems to be the visible telephone system in its entirety, as inserted within various architectural situations. More than a mere technical manual, then, this second version of the booklet reveals itself to be a propagandistic document that introduces the aesthetic dimension of telecommunications facilities. "Such permanent built-in features have the double advantage of being both useful and decorative." 8 Once again, the authors insist on the adaptability of the service to different styles of housing, from the very modest one-story bungalow to the elaborate residence that is best served "by a complete intercommunicating service in addition to

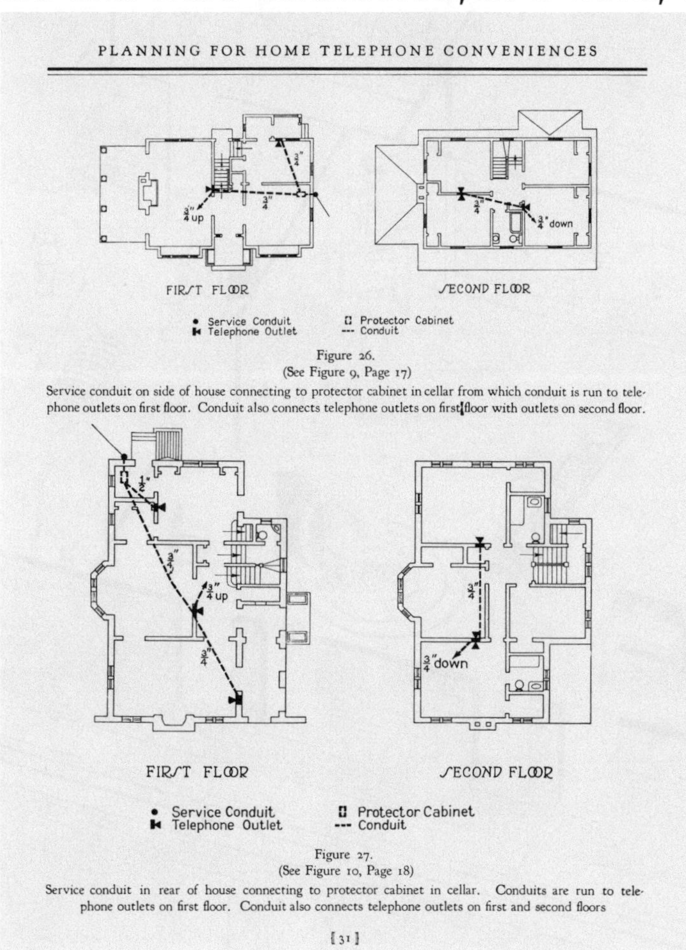

5 *Modern Telephone Service for the Home* (American Telegraph and Telephone Company, 1929), n.p.

6 Jonathan Sterne, "Sound Out of Time/Modernity's Echo," in Carol A. Stabile (ed.), *Turning the Century: Essays in Media and Cultural Studies* (Boulder, CO: Westview Press, 2000), pp. 9–30.

f.4 *Planning for Home Telephone Conveniences*, AT&T, 1928.

7 Adrian Forty, *Objects of Desire: Design & Society from Wedgwood to IBM* (New York: Pantheon Books, 1986). See in particular chapter 8 "Electricity—The Fuel of the Future," pp. 182–206.

8 *Modern Telephone Service* (see note 5).

f.5 *Modern Telephone Service for the Home,* AT&T, 1920s.

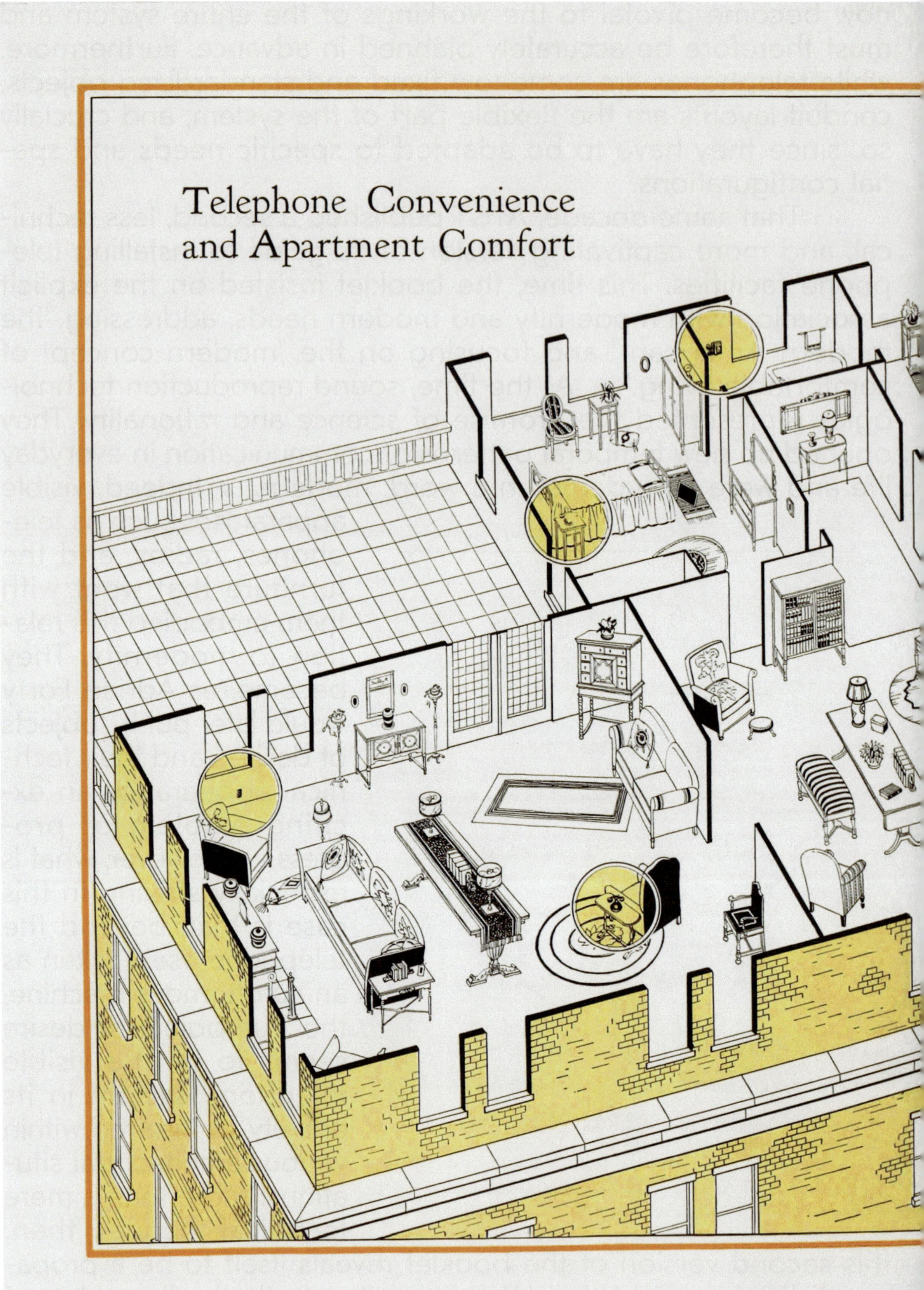

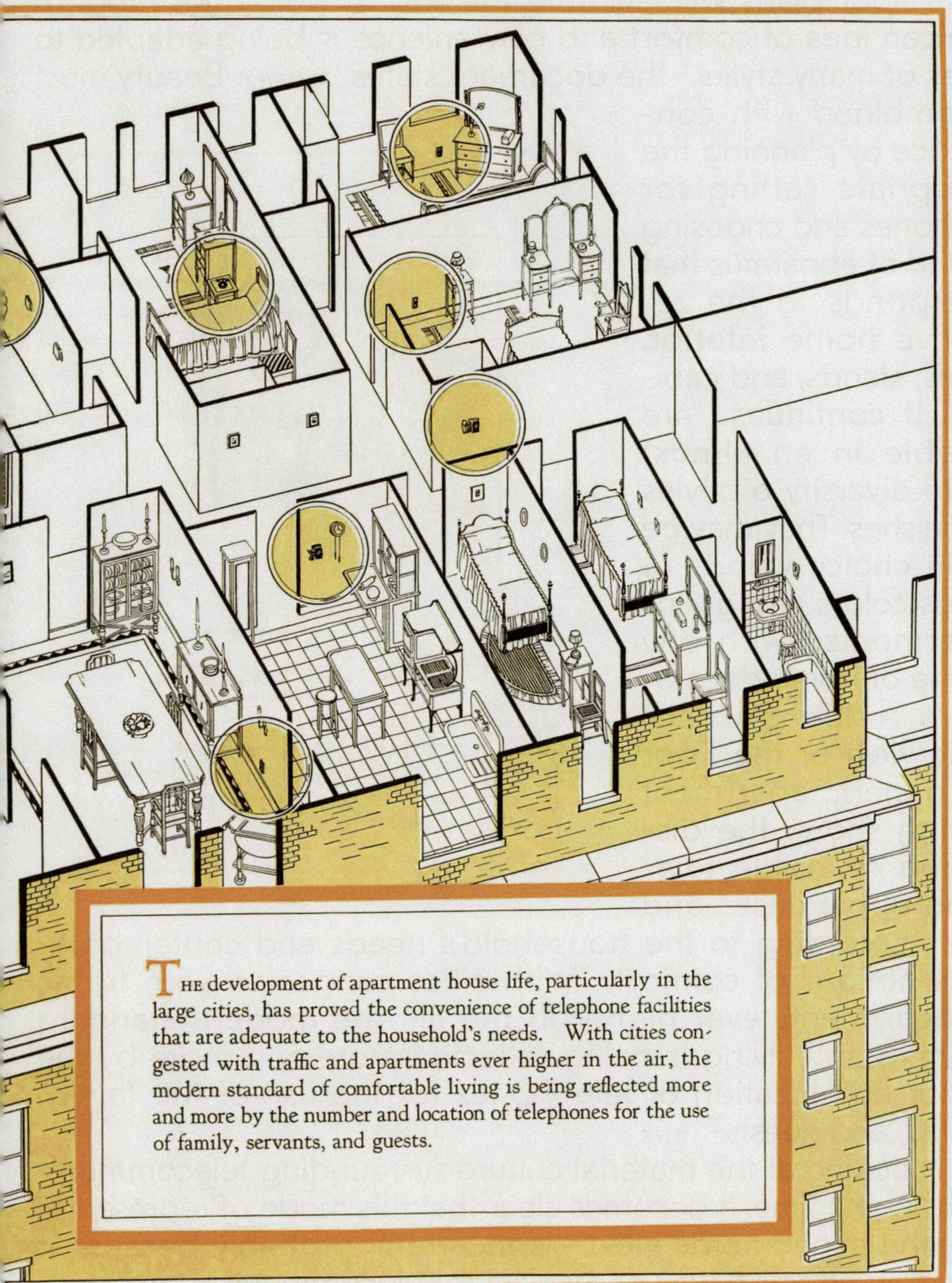
The development of apartment house life, particularly in the large cities, has proved the convenience of telephone facilities that are adequate to the household's needs. With cities congested with traffic and apartments ever higher in the air, the modern standard of comfortable living is being reflected more and more by the number and location of telephones for the use of family, servants, and guests.

the ordinary facilities." 9 The same series of plans related to various typologies of home is displayed by way of example: "A few typical floor plans will illustrate the way in which the modern American idea of comfort and convenience is being adapted to homes of many styles," the document states. 10/f.6/f.7 Beauty must be combined with convenience by planning the appropriate setting for telephones and choosing the kind of apparatus that corresponds to the respective home interior. "Tables, stands, and cabinets," it continues, "are available in an almost infinite diversity of styles and finishes. They may be had in choice woods of various colors and grains to harmonize with any scheme of home decoration." 11 A striking axonometric view of the floor of a generic apartment building shows the disposition of visible telephone apparatuses and outlets according to the household's needs and contemporaneous notions of comfort. "With cities congested with traffic and apartments ever higher in the air, the modern standard of comfortable living is being reflected more and more by the number and location of telephones for the use of the family, servants, and guests." 12/f.5

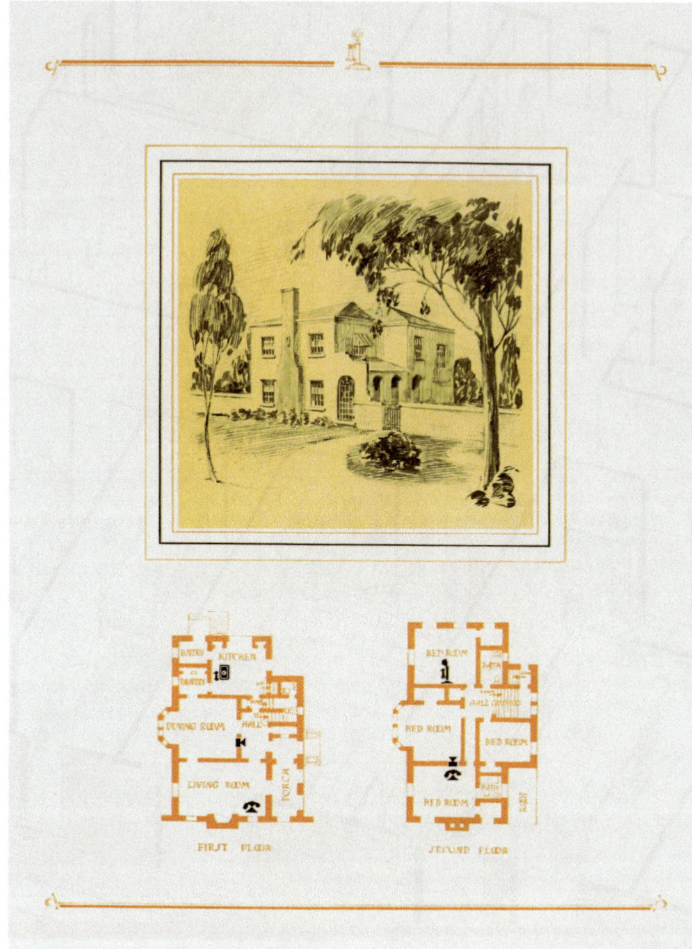

Looking at the material culture surrounding telecommunications at the time, it becomes clear that this mode of representation—the axonometric view—is recurrent, probably because it offered an effective way of displaying telecommunications facilities and their integration within architecture, in particular within architectural planning. Another, more synthetic booklet for telephone installation presents an axonometric view of two empty generic homes—a small or medium-sized one, and a larger one—which are both equipped with what the booklet calls "typical telephone conduit layouts". 13/f.8/f.9 Again, the authors' ultimate message is that, in order to buy or build a modern home, one needs to anticipate the installation of telephone conduits

9 Ibid.

10 Ibid.

f.6 *Modern Telephone Service for the Home*, AT&T, 1920s.

11 Ibid.

12 Ibid.

13 *Home Telephone Conveniences* (see note 2), p. 29.

during the planning process. These stylized versions of axonometric views, bare of interior furniture and other domestic equipment, further emphasize the essential need to couple telephone facilities with primary elements of construction such as walls, floors, and the roof.

As Yve-Alain Bois and previously Bruno Reichlin observed, axonometry is a mode of representation that became widespread during the modern period. 14 Prior to that time, and since the Renaissance period, the illusionistic perspective posited by Alberti and Piero della Francesca had been the privileged mode of representation, reflecting cultural models of the epoch. 15 Perspective focuses on the observer's perception of an object from a given spatial position, whereas axonometry preserves in scale the given object's actual dimensions, independently of its spatial location. Since the Middle Ages, axonometric projection had been used for technical drawings and deployed in the construction of fortifications and machines, whenever detail, precision, and the workings of the objects concerned was a primary practical necessity. 16 From the early twentieth century onwards, the use of axonometry became more widespread, extending way beyond the specialized, technical fields.

f.7 *Modern Telephone Service for the Home*, AT&T, 1920s.

14 See Yve-Alain Bois, "Metamorphosis of Axonometry," *Daidalos*, 1 (1981), pp. 40–58, and Bruno Reichlin, "The Axonometric as a Project: A Study of Alberto Sartoris," *Lotus International*, 22 (1979), pp. 83–93.

15 On the cultural use of axonometry throughout Western history, see Massimo Scolari, *Il disegno obliquo: una storia dell'antiprospettiva* (Venice: Marsilio Editori, 2005). In particular, see the chapter "Disegnare in *paralleli modo*," pp. 284–93.

16 See chapters "Machinationes" and "La prospettiva soldatesca," in ibid., pp. 229–84.

The meaning of this change is threefold. Firstly, this mode of representation turns out to be the expression of a new functionalist regime that prompts a major cultural shift, from a conception of architecture as a static and formalized object to one that calls attention to its performative and mechanized processes. Secondly, unlike the classical perspective, which limits the observer to a single viewpoint situated outside the building itself, the axonometric view allows observers to enter the volume in order to gain a thorough picture of it. And lastly, from that inner point of view, the user becomes a sort of ubiquitous figure, potentially able to live (and communicate) in every part of a house, from the bathroom to the living room, and so on. In providing not a single, central viewpoint but an ensemble of equitable views, axonometry suggests a temporal movement through a building's various interiors and produces thus a "cinematic" perception of architecture. 17 Modern architects seeking to destroy the traditional supremacy of the façade that was inherent to classicism found in axonometry a suitable way to represent their buildings from multiple perspectives, and to draw attention to the spatial and structural features of their work. Due to its ability

17 See Yve-Alain Bois, "Montage and Architecture," *Assemblage*, 10 (1989), pp. 111–15, in particular on Auguste Choisy's cinematic use of axonometry.

f.8 Advertising brochure for AT&T, date unknown.

to represent the third dimension of architecture, axonometry also facilitates the display of different layers or phases of construction. Compared to a fixed and finished perspective, it tells the observer more about the planning process by revealing, for instance, the presence of conduits buried deep within a structure's walls and the different technical apparatuses that occupy architectural interiors. Thus, this mode of representation makes visible modern architects' interest in technology since it reveals the functioning of buildings, positing them as whole, spatially and technically unified systems.

Another telecommunications document from the George Clark Radioana Collection (at the Smithsonian Institution) shows the ensemble of conduits powering an array of radio devices distributed throughout each room of the home, with the exception of the stairs, by way of a perspective sectional drawing. f.10 Electrical

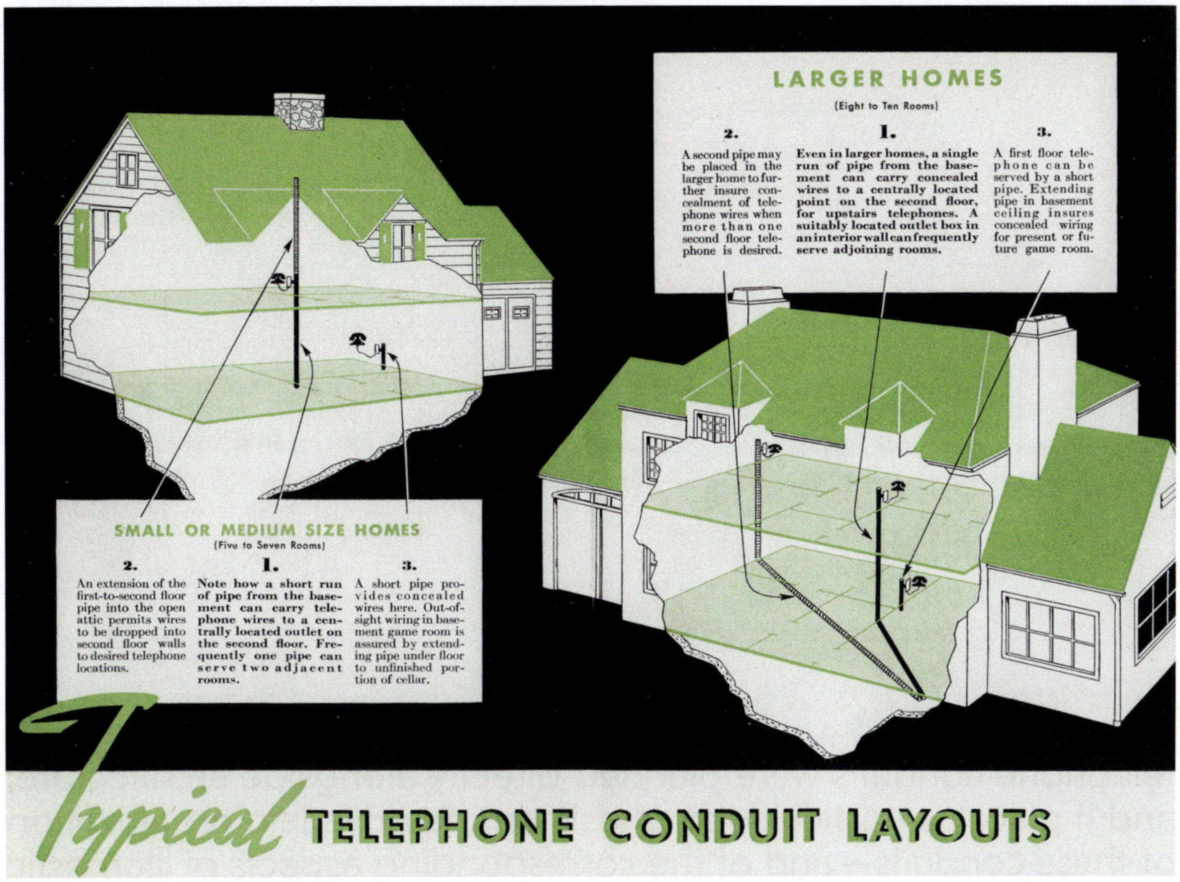

f.9 Advertising brochure for AT&T, date unknown.

conduits, concealed within the walls and floors, enable different kinds of equipment to produce sounds and music addressed to a new type of domestic audience. As a typical product of the culture of the nineteenth century, this kind of analytical section displays the "interior as a façade," thereby coupling the classical perspective that keeps the observer at a distance, outside, with the modernist need to represent technical services hidden within the architecture. [18] In thus highlighting by means of graphic representation that latter tension between architecture and technics, and architecture and construction, the document serves not only to corroborate a progressive interest in architectural representation on the part of telecommunications propaganda but also incisively illustrates a gradual transition to modernity. [19]

To conclude, then, from the late 1920s onwards, documents from telecommunications companies seemed to borrow directly from the architectural culture of the epoch. The broad use of plans and axonometries in advertising materials attests both the intent to address professionals in a direct manner and the endeavor to make the technical language of architecture available to the general public. f.11/f.12/f.13/f.14 Like technology, technics—or rather, the art of fabricating things—was now no longer something to be hidden, but an exciting marker of progress, something that must be put on display, emphasized, and even aestheticized. Instead of calling attention to the latest sound reproduction technology—the telephone in this case—this series of images highlights that

18 Georges Teyssot, "A Topology of Thresholds," in *Home Culture*, 2, no. 1 (2005), pp. 89–116.

19 Antoine Picon, "Du traité à la revue, l'image d'architecture au siècle de l'industrie," in Stéphane Michaud, Jean-Yves Mollier, and Nicole Savy (eds.), *Usage de l'image au XIXe siècle* (Paris: Créaphis, 1992), pp. 153–65.

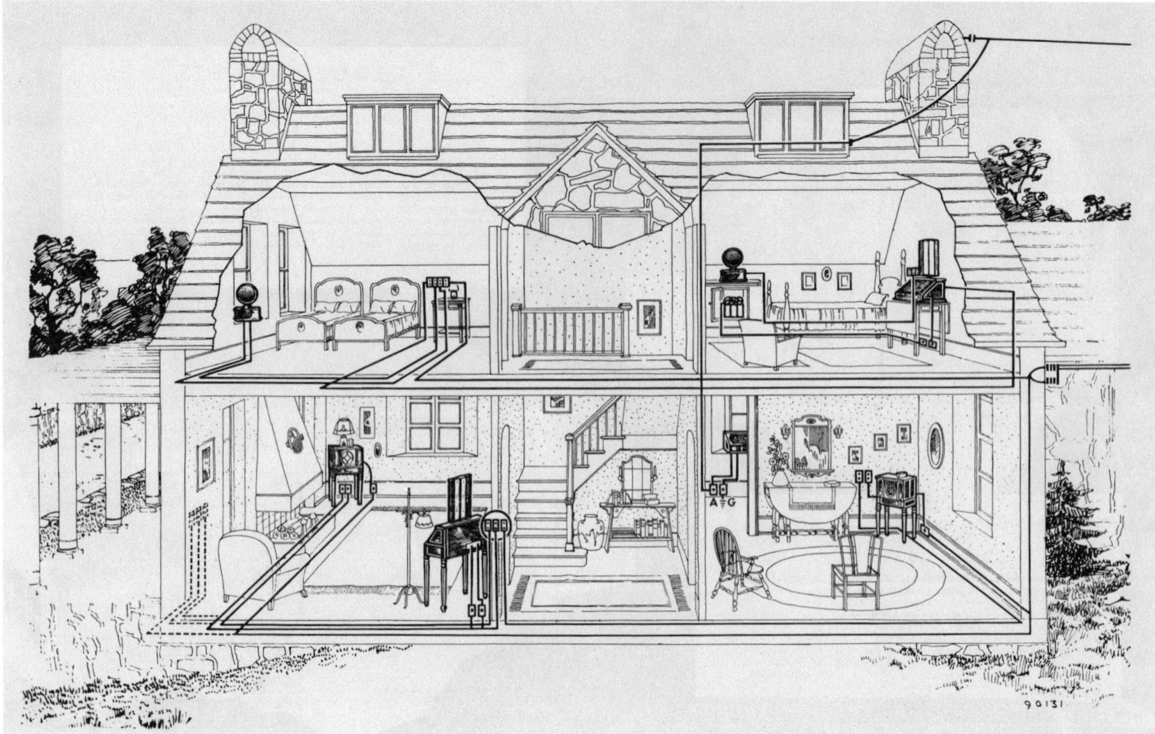

f.10 Drawing from the George H. Clark Radioana Collection, date unknown.

telephone conduits were planned directly alongside architecture; and it places a particular emphasis on the articulate organization of these conduits—and of the corresponding aspects of domestic life—as rendered in architectural plans or axonometrical views. Generally speaking, drawings function as mediators of architecture since they convey select information about an actual object. In selecting a specific graphic language, one chooses which particular content must be maintained during the transition from architecture itself (the object) to its mediator (the drawing). [20] Thus, the ensemble of documents produced as part of telecommunications propaganda materials must be seen as a typical expression of the agenda of the so-called machine age, one which voluntarily stresses the technicality of things.

f.11 Advert for AT&T, 1947.

From among the variety of interpretations of modernity to be noted at the time, two diametrically opposed positions

20 Bruno Latour, "How to Be Iconophilic in Art, Science and Religion?" in Carrie A. Jones and Peter Gallison (eds.), *Picturing Science, Producing Art* (London: Routledge, 1998), pp. 418–40.

f.12 Advert for AT&T, exact date unknown [193?].

emerged in regard to machines. On the one hand, the sublimation of finished and geometrical objects as plastic products with inherent symbolic values. This is a formal position that is given full expression, for instance, in the 1934 Machine Art exhibition at MoMA. In his introduction to the catalog, Philip Johnson states that the "beauty of machine art is in part the abstract beauty of 'straight lines and circles' made into actual tangible 'surfaces and solids' by means of tools, 'lathes and rulers and squares.' … Machines are, visually speaking, a practical application of geometry." 21 This geometric infatuation led him, along with Alfred H. Barr Jr., to curate an exhibition on modern design that combined tools with machines, from the spoon to the microscope, without making any proper distinction between them and without questioning how they functioned. On the other hand, the operative behavior of the machine is central to the emergence of a more processual viewpoint. This position is well represented by another modern thinker: indeed, the very same year as the MoMA exhibition, Lewis Mumford published *Technics and Civilization*, wherein he defined the machine as "the ensemble of knowledge and skills and arts implicated in the new technics." 22 The degree of a machine's independence from its operator—its automatic nature—and the degree of specialization within its functions are, according to Mumford, the constituents of the fundamental difference between a machine and a tool.

21 Philip Johnson and Alfred H. Barr Jr., *Machine Art* (New York: The Museum of Modern Art, 1934), p. 3.

f.13 Advert for AT&T, 1947.

22 Lewis Mumford, *Technics and Civilization* (New York: Harcourt, Brace & World, 1963 [1934]), p. 12.

These two positions can quite easily be set into a direct dialogue with the contemporary material culture of telecommunication. The first understanding then relates to representations that rhetorically as well as visually place the sounding end device—the telephone, the radio, etc.—at center stage. The latter definition, which insists on the automation of functions and the specialization of processes, in contrast relates to representations that emphasize the underlying technical systems and the spatial and temporal relations and practices involved—just as the specific corpus of documents taken from telecommunications companies' propaganda materials and described above does. In this sense, I would suggest that these representations of conduits within architectural interiors offered a machinic reading of telecommunications and, accordingly, of architecture as well. More than singular sound reproduction technologies and isolated objects, the systems of sound conduits ultimately reveal the processes and workings of a new era of domestic communication technologies.

f.14 Advert for AT&T, 1938/39.

Tapping Machines: Listening to Difference, 1928–1956

Sabine von Fischer

Sabine von Fischer is Research Associate at the Zurich University of Applied Science, School of Architecture, Design and Civil Engineering.

Privacy and concerns about noise partake in the history of architecture, science, technology, society, industry, and modernity. Testing architecture's capacity to grant privacy is a task at the nexus of objective measurement, environment, perception, construction, and comfort. Since the 1920s, the coincidence of new construction methods in housing production and distress over privacy challenged the modern architect's belief that lighter buildings with thinner walls would pave the way to the future. This essay examines testing methods devised for measuring sound transmission through the ceilings and floors of neighboring apartments. It discusses sound insulation in architecture as a technological momentum for fulfilling architecture's function of securing acoustic privacy, and addresses difference in multiple ways. It refers to the spatial situations where sound measurements were taken, namely above and below a floor. Then it addresses conflicts between automated measurements and techniques of listening. In short, it aims to deliver another and possibly different history of objectivity in the modern period in architecture.

One of the instruments created in the late 1920s, in an endeavor to resolve the problem of inherent subjectivity in sound assessment, was the "machine for producing impact sounds." Later it became known as the "tapping machine," or "Hammerwerk" in German, or "machine à chocs" in French: a stunningly simple mechanism that hammered on floors to test their acoustic qualities as well as the level of sound insulation provided by various construction types. This hammering apparatus appears at first glance to bring architecture and machine into a straightforward sonic relationship. However, the human hearing threshold, until around 1930 was part of the test too, and this entangled architecture and machine in a more complicated relationship, one of mediatized sensation. Automation of the tests relieved the acoustic sciences of this tension. Yet still, the differences between measuring sound with an apparatus and assessing sound by hearing left ample room for debate.

Designed to test construction methods in the laboratory as well as on-site, the tapping machine also tested the test setting itself. Tackling this problem of the complicated relationship between hammering, acoustic measurement, sonic perception, and privacy as a function of architecture calls for a "*stubbornly realist attitude*." [1] Such a realist attitude is an aid to overcoming the self-confinement of architectural criticism in formal or aesthetic loops. In the history of twentieth-century architecture,

1 Bruno Latour, "Why Has Critique Run Out of Steam? From Matters of Fact to Matters of Concern," *Critical Inquiry*, 30, no. 2 (2004), pp. 225–48; here p. 231. Emphasis in original.

one of the few to address comparable links between technology and comfort was Reyner Banham, who also warned that, "like domestic heating systems, social control systems are disappearing into the floor slab." 2 The pipes of such systems transmitted heat as well as unwelcome sounds into neighboring spaces, and remained invisible but not unheard. Technical apparatuses, laboratory practices, questions of physical comfort, and complaints about noise are a complex that architectural historiography has yet to deal with in full. The history of the tapping machine offers an opportunity to begin such an investigation.

2 Reyner Banham, "Softer Hardware," *Ark*, 44 (1969), pp. 2–11.

The first part of this paper presents a brief history of the tapping machine. I explain the mechanics and the uses of the apparatus, the design of which was published for the first time by the U.S. National Bureau of Standards in 1928, and subsequently in modified form by other institutes on both sides of the Atlantic. The second part tells of a 1956 episode of reintroducing human hearing into the testing of architectural soundproofing. It questions the master narrative of modern standardization. Long after automation had become the norm for acoustic measurements, Germany's leading expert in architectural acoustics devised a combination of the standardized method for assessing impact sound and listening as a psychophysiological process, when he designed a simple method which included the human ear. Based on the difference between the perceived loudness of two standardized impulses inside and outside of a room, his method provided a practical tool for measurements on-site, beyond the laboratory walls.

Sleepless in the metropolis, c. 1930

"Wie kann er schlafen durch die dünne Wand?" 3 This refrain—"How can he sleep behind a wall so thin?"—was quoted from an earworm hit in a bestselling novel by the former journalist Gabriele Tergit. The novel tells of life in Berlin around 1930, of mass housing, speculation and corruption in the building industry, and of the rise to fame of a songwriter who gains citywide acclaim with this one song in which noise anxiety plays a key role. Changes in lifestyle clashed with changes in architecture, when a growing variety of noise sources sounded inside of modern, less noise-absorbent, and more densely populated building types. Steel framing, prefabrication, mass production, plumbing and wiring for every unit and every room, and the proliferation of domestic appliances had acoustic consequences, particularly in combination with the increasing density of cities with multi-unit and multistory dwellings. The walls and floors of such dwellings

3 Gabriele Tergit [= Elise Reifenberg], *Käsebier erobert den Kurfürstendamm* (Berlin: Das Neue Berlin, 2004 [1931]), p. 149. For a cultural history of the architectural implications of the novel, see Ines Lauffer, *Poetik des Privatraums: Der architektonische Wohndiskurs in den Romanen der Neuen Sachlichkeit* (Bielefeld: transcript, 2011), pp. 83–117.

transmitted a multitude of bothersome noises to neighboring apartments. f.1

Practices as well as debates concerning sound in the built environment indicate that acoustics were increasingly perceived as integral to the program and function of architecture from the 1920s onwards: while modernist architectural discourses in *L'Esprit Nouveau* and other publications discussed musical techniques and acoustic ambience, the acoustic performance of buildings came to be included in a new scientific apparatus by which architecture was to be planned, built, and evaluated. Architects, scientists, the building industry, and society at large were no longer able to rely solely on empirical knowledge handed down over generations but instead had to tackle questions of architectural performance by examining a multitude of functions, for each of which a set of tools was designed that could assess their feasibility, as part of modernity's endeavor to quantify architectural parameters.

f.1 One man's music is another man's noise (1934).

Acoustics, alongside other new considerations rooted in building physics and climate control (such as lighting, temperature, humidity, and hygiene in general) changed the ways in which structures were described in the early twentieth century. Architecture was seen not only to require a three-dimensional design, but also to involve temporal and programmatic scenarios. John Summerson defended this notion in 1957 by rejecting the "axiom that architecture is an affair of simple geometric forms," and including "the spatial dimensions, spatial relationships and other physical conditions required for the convenient performance of specific functions" in the problems of architectural theory. 4 Summerson's term "performance," with its emphasis on dynamic processes, has proven a useful means to circumvent naïve, static, and deterministic notions of function.

For historiography, function provides a crucial hinge between technological, economic, and social histories. It has been discussed in many different contexts, often in reference to Louis Sullivan's influential plea of 1924, for a utility wherein "forms ... would grow naturally out of ... needs." 5 Notwithstanding the vehement rejection of functional determinism by the cybernetic

4 Summerson, in his original contribution to the quest to find unity in modernity's diverging concepts, notes Bruno Zevi's conception of the organic as a key moment in architectural theory's move beyond formal abstraction, and an inclusion of the social sphere. John Summerson, "The Case for a Theory of Modern Architecture," *The Journal of the Royal Institute of British Architects*, 64 (1957), pp. 307–13; here pp. 308, 309.

5 Louis H. Sullivan, *The Autobiography of an Idea* (New York: Dover, 1956 [1924]), p. 258. Adrian Forty in his comprehensive chapter on the genealogy of the word "function" in architecture links Sullivan's notion to the German term "Zweck," equivalent to purpose or destiny. Adrian Forty, "Function," in Forty, *Words and Buildings: A Vocabulary of Modern Architecture* (London: Thames & Hudson, 2004), pp. 181–7. For another discussion of the numerous and diverse definitions of "function," see, for example, Ute Poerschke, *Funktionen und Formen: Architekturtheorie der Moderne* (Bielefeld: transcript, 2014).

and hippie generations alike, function is inherent to all architecture. For one, Julius Posener in his "critique of the critique of Functionalism," posited that the striving for functionality partakes in the history of social reform. [6] It is along these lines that the necessity of linking scientific, technological, and social histories of architecture has to be addressed with a "*stubbornly realist attitude*," as stated earlier in this essay.

The technological problem of noise and soundproofing is closely bound up with the social histories of housing and cities. In the late nineteenth century, noise was considered the second greatest threat to health, after odors. [7] Noise complaints increased—on both sides of the Atlantic. [8] Amid all this noise about noise, some historians tell the story that sensitivity to bothersome sounds shifted from being a bourgeois to a mass phenomenon, and thus proliferated. [9] Aspirations to privacy certainly increased, and were further invigorated by the construction industry's market launch of a growing number of insulation products. Counting complaints might give us quantitative evidence and sustain the story of raised sensitivity, yet there was no objectified method for measuring actual noise until the late 1920s. There were "noise units," "sensation units," and "transmission units," but not yet a quantified authoritative measure to say what was too loud. With units such as the Phon, introduced in Germany in 1925, and the deciBel, introduced in the U.S. in December 1928, objective measurements independent of the subjectivity and inconsistency of human perception were now at hand, and they successively paved the way for a standardized system of sound assessment. [10]

In parallel to finding an objective reference for sound intensity, a constant and controllable impulse for what was measured was needed. Thus, for testing air-borne sound (sound waves travelling through the air), sine waves of different pitches were played from gramophones. The more difficult problem was assessing impact sound (sound waves travelling through solid matter, such as a building structure, and especially the floors and the airspace below), as in the case of a neighbor's footsteps.

In 1928, acoustic specialists Vivian L. Chrisler and Wilbert F. Snyder of Washington D.C.'s National Bureau of Standards introduced a new apparatus in a research paper. This "machine for producing impact sounds" hammered steadily on the floor and thus could serve as an objective reference in various situations. [11] Chrisler, in a subsequent paper, pointed out that, "In many ways floor structures present a particularly difficult problem, as we have to deal in this case not only with air-borne noises but with impacts produced by walking or by moving furniture." [12]

6 Julius Posener, "Kritik der Kritik des Funktionalismus," *ARCH+*, 7, no. 27 (1975), pp. 11–18, here pp. 17, 15. Many thanks to Nikolaus Kuhnert, who introduced this text into my seminar's discussions of architectural criticism at the ETH Zurich in 2012.

7 Geneviève Massard-Guilbaud, *Histoire de la pollution industrielle: France, 1789–1914* (Paris: EHESS, 2010).

8 See Emily Thompson, *The Soundscape of Modernity: Architectural Acoustics and the Culture of Listening in America, 1900–1933* (Cambridge, MA: MIT Press, 2002), pp. 115–68; Karin Bijsterveld,. *Mechanical Sound: Technology, Culture, and Public Problems of Noise in the Twentieth Century* (Cambridge, MA: MIT Press, 2008), pp. 159–92.

9 See Bijsterveld, *Mechanical Sound* (see note 8), p. 238.

10 Ibid., pp. 104–5; Thompson, *The Soundscape of Modernity* (see note 8), p. 158.

11 V. L. Chrisler and W. F. Snyder, "Transmission of Sound Through Wall and Floor Structures," *Bureau of Standards Journal of Research*, 2 (1928), pp. 541–59.

12 V. L. Chrisler, "Measurement of Sound Transmission," *Journal of the Acoustical Society of America*, 1, no. 2A (1930), pp. 175–80; here p. 175.

f.2 a–c "Machine for producing impact sounds" at the National Bureau of Standards in Washington D.C. (1928); the "Hammerwerk" at the Technical University Berlin (1936); the heaviest tapping machine documented at the Institution of Civil Engineers in London (1946).

Interrupting the flow of sound through a building's structure called for new concepts in engineering materials and joints. Testing sound transmission between rooms and through floors was a problem that Chrisler and Snyder would resolve: their machine "for experimenting with impact noises" featured five hammers of 500 grams each, lifted by cams and then dropped at intervals of about one-fifth of a second. 13 This weight and speed proved most suitable, although other laboratories built numerous alternatives.

A German tapping machine documented eight years later hammered at a speed of ten times per second, thus at double the speed of its American predecessor. 14 British engineers devised a tapping machine that assured 7.33 impacts per second. Here, as in the German apparatus, the hammers were arranged in a circle, yet with hammerheads of two kilograms each, making this the heaviest machine documented in the handbooks and literature consulted during my research. 15 Meanwhile, the tapping machine conceived in Goteborg, Sweden, hammered only twice per second. 16/f.2 a–c

The National Bureau of Standard's initial setting of weight and speed later became the reference for other nation's standards of impact sound measurement, including Germany's institutionalized standard DIN 52210 and, in the process of international coordination, the international standard for building acoustics ISO 140. 17

Today, standards internationally follow the rhythm of the American machine, but neither this nor any other of the apparatuses documented in laboratory papers disseminated in the postwar period can be heard or seen today, as none of them were deemed worthy of archival conservation. We are left with data describing the rhythms of their tapping, and with dates indicating the moment in time when building physics became a part of construction bureaucracies, at a multisonorous piece-rate, before their standardization. It is the noise of the many tapping machines in this narrative that draws attention to the practices brought about by regulation of the design, the construction, and the sonic performance of buildings.

13 Ibid., p. 179.

14 A. Gastell, "Schalldämmungen in der Praxis und Vorschläge zur Normung des Schallschutzes," *Akustische Zeitschrift*, 1, no. 1 (1936), pp. 24–35; here p. 29; Arnold Schoch, *Die physikalischen und technischen Grundlagen der Schalldämmung im Bauwesen* (Leipzig: Hirzel, 1937), p. 104.

15 N. Fleming and W. A. Allen, *Modern Theory and Practice in Building Acoustics* (London: The Institution of Civil Engineers, 1946), referenced in Leo L. Beranek, *Acoustic Measurements* (New York: Wiley, 1949), p. 886.

16 Per V. Brüel and Harry K. Zaveri, "Of Acoustics and Instruments: Memoirs of a Danish Pioneer – Part 2," *Sound&Vibration* (August 2008), pp. 14–32; here p. 16.

17 For DIN 52210 ("Einheitliche Mitteilung und Bewertung von Messergebnissen") by the Deutsches Institut für Normung. For ISO 140 ("Acoustics – Measurement of Sound Insulation in Buildings and of Building Elements") by the International Standards Organization.

Steel-construction sound testing, 1936

In 1936, a group of German scientists left the laboratory to test the acoustic insulation inside a low-rise and a high-rise apartment building at the Reichsforschungssiedlung Spandau-Haselhorst. They were equipped with their equivalent of the American tapping machine: an apparatus called "Hammerwerk." 18 Haselhorst was the largest housing complex built under the Weimar Republic, and also a large-scale construction research project. 19 Between 1931 and 1935, 3,500 dwellings were completed. Various materials and techniques were used to build the long rows of housing in the complex, the intention being to compare these at various stages, at full scale, also once inhabited and in use. Yet the Nazis' seizure of power in 1933 overshadowed Haselhorst's completion, put an end to modernism, and changed the lives of its inhabitants as well as preferences in architectural form (to the extent that pitched roofs were added in the final phase), although not the modern project of assessing and standardizing architecture's performance. The research program of testing economical construction methods was pursued throughout the 1930s, and steel frames continued to be used in some of the multi- or single-story rows of housing, also in the project's final phase. 20/f.3 Steel frames were known to be particularly problematic for the transmission of impact sound from one unit to another in apartment buildings, not only to those adjacent but also to dozens of others nearby; and so, despite being both extremely effective in structural terms and low in cost, they had to be paid close attention when it came to acoustic testing. At Haselhorst, cork sheets of 5 mm were layered between the steel frame and the brick infill, and the efficiency of this intervention was evaluated in the tests of 1936. 21

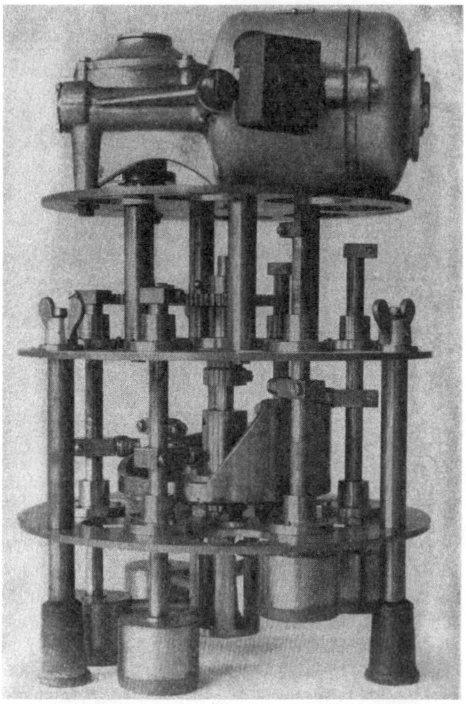

From the very start, the Haselhorst experiment proved arduous for everyone involved. It was time-consuming to obtain all the residents' permission to carry out noisy experiments in

18 Gastell, "Schalldämmungen in der Praxis" (see note 14), p. 29.

19 After disputes with Walter Gropius and Stephan Fischer, who had initially won the competition, the planning was put into the hands of Paul Mebes, Otto Bartning, and Wilhelm Lübbert. In order to loosen the rigidity of Gropius's initial plan, the rows were shifted slightly off the grid and accentuated by L-shaped building heads. See Michael Bienert, *Moderne Baukunst in Haselhorst: Geschichte, Bewohner und Sanierung der Reichsforschungssiedlung in Berlin-Spandau* (Berlin: Berlin Story, 2015), pp. 31–4. Since the *Siedlung* was modern and completed after the Nazi takeover, it was never documented but rather abandoned to historical oblivion until recently, when cultural historian Michael Bienert was commissioned by the housing cooperative GEWOBAG to pull the materials from the archive. Bienert was so kind as to give me a tour of the complex, including the apartment now maintained as a "museum apartment," in August 2015.

f.3 Steel frame construction at the Siedlung Spandau-Haselhorst, acoustically tested with the "Hammerwerk" in 1936; in the background, traditional brick construction (1930).

20 Ibid., pp. 42, 67.

21 Gastell, "Schalldämmungen in der Praxis" (see note 14), fig. 11. Two years later, a similar test was documented for Alfred and Emil Roth and Marcel Breuer's Doldertal buildings in Zurich. See E. Pestalozzi, "Schallschutz im Hochbau," *Schweizerische Bauzeitung*, 111, no. 9 (1938), pp. 108–10.

their apartments. The high-pitch test tones sent through the walls hurt the residents' ears, and the hammering noises resounded throughout the building. 22 Often, the tapping machine could not be placed in the center of a room because a bed or other furniture was in the way. Especially in bedrooms, the level of sound absorption by carpets, curtains, wall coverings, bedding, and upholstery varied greatly—so much so that the results were hard to interpret, and the enormous effort of leaving the laboratory must have appeared questionable to the scientists. 23

The measurements taken at Haselhorst and in other buildings in Berlin are of twofold interest in the history of architecture. Firstly, the various construction methods used there were part of the larger experiment of using steel in multistory housing construction. Secondly, the measurements were taken as test cases, to establish reference values in the standardization of tests. When new construction methods were filed for building permits, the measurements from the 1936 experiments served first as a reference and were then adopted in the second edition of "Technical Specifications for the Approval of New Construction Methods" of 1938, the DIN 4110. 24 Here, for the first time in German building regulation, the impact sound level was defined in numerical quantities, namely it was not to exceed 85 Phon. 25 The acoustic testing prescribed by the DIN regulations referred to laboratory measurements; these numerical values, however, were not released before having been tested outside of the laboratory, at Haselhorst and in other buildings in the Berlin area, at full scale, and with all the imprecisions and mishaps in planning, production, and implementation in play.

f.4 a–b Portable tapping machine for use outside of the laboratory (1952), and domesticated in the living room (1960).

Some early tapping machines could hardly be lifted, let alone be moved around in multistory buildings, outside of the laboratory. Also, housing projects under construction were usually not wired for electricity, which many such apparatuses relied on. With the growing interest in soundproofing, portable models were built. These often came with a box or case, even with a handle. Scientists now wore their clean-room suits also in the midst of homely ornament and floral carpets, and by the 1950s, the tapping machines in technical handbook illustrations appeared thoroughly domesticated. f.4 a–b In 1950, the Danish firm Brüel & Kjaer launched the first standardized tapping machine, which was nicknamed "Normtrampler" (standardized tramper) in a technical

22 Gastell, "Schalldämmungen in der Praxis" (see 14), pp. 32–3.

23 Ibid., p. 31.

24 DIN 4110 ("Technische Bestimmungen für Zulassung neuer Bauweisen") by the Deutscher Normenausschuss (with a circular by the Reich Minister from July 12, 1938), in *Zentralblatt der Bauverwaltung vereinigt mit Zeitschrift für Bauwesen*, 58, no. 32 (1938), pp. 879–87. The first edition of DIN 4110 by the Deutscher Normenausschuss (German Standards Committee) was published in *Zentralblatt der Bauverwaltung vereinigt mit Zeitschrift für Bauwesen mit Nachrichten der Reichs- und Staatsbehörden*, 54, no. 9 (1934), pp. 563–8. Here, soundproofing was defined relative to common brick walls, without numerical values. For a longer discussion of acoustic regulation in Germany and Switzerland, see Sabine von Fischer, "Dynamique ou uniformisation? À propos de la normalisation dans la conception de projets et l'industrie du bâtiment, à travers l'exemple de la réglementation acoustique," *Matières*, 12 (2015), pp. 116–31.

25 DIN 4110: 1938 (see note 24), p. 887.

handbook. 26 Its domestic-looking wooden casing, a design still in use today, also falls into the category of those "archaic objects" often used to reconcile new technology with the familiar. 27 At the same time, the human activity of loud footsteps was translated into mechanical movements and technical standards, a kind of early sound-making yet non-hearing man-machine. Acoustic devices for architectural testing were designed with the utmost possible exclusion of experience in mind. These loud, penetrating devices reduced sound to a physical impulse stripped of any variation in pitch, timbre, and rhythm. The only intended recipient of these sounds was the measuring apparatus, which after 1930 was no longer the human ear.

The logic of the laboratory posed further problems in the world outside, where the tapping machine was taken as a universal reference in assessments of the transmission of (mostly unwelcome) sounds. These unchanging hammering sounds differed from the widely diverse noises likely to cause conflicts between neighbors, to say nothing of the many different types of footstep, from bouncing heels to scuttling toes. Thus, the tapping machine was repeatedly compared to different types of male or female, soft or hard-heeled footsteps, and Karl Gösele, the leading German expert on impact sound transmission, concluded in his research report of 1957 that some flooring materials performed better in respect to their soundproofing capacity when tested with standardized hammer heads, others when walked on by real people. 28

Considering that the sole of a shoe and the mood of a person altered the quality of impact sounds, what about screeching chairs, rattling dishwashers, droning vacuum cleaners, and plummeting children's toys? Despite, and equally because of this reduction of the world of noises to a standard of five hammer heads of 500 grams each, the tapping machine's beating in a 1967 study was deduced to be a "very practicable compromise, ... because sounds are not only produced by walking and running, but also by knocking, and when things are dropped." 29

26 Werner Bürck, *Die Schallmessfibel für die Lärmbekämpfung* (Mindelheim: Sachon, 1955), p. 64.

27 Adrian Forty coined the term "archaic objects" for radios concealed within antique furniture in the 1920s as a "resistance to the newness of things." Adrian Forty, *Objects of Desire: Design and Society since 1750* (London: Thames & Hudson, 1986), p. 11.

28 Karl Gösele, *Zur Messung und Beurteilung des Trittschallschutzes* (Stuttgart: Forschungsgemeinschaft Bauen und Wohnen, 1959), p. 26.

29 L. Cremer and M. Heckl, *Körperschall: Physikalische Grundlagen und technische Anwendungen* (Berlin: Springer, 1967), p. 295 (My translation).

The ear revisited, 1956

Concerns in architectural acoustics were deeply entangled with the tasks of granting privacy and enabling communication in dwellings and workplaces. Testing, assessing, and comparing the sonic performance of buildings relied in part on quantification of the residents' acoustic experience of these dwellings and workplaces. Thus, acoustic testing and standardization came with a certain tension between method and aim: while the aim was to improve the comfort of living and working, the methods themselves excluded human experience. The human presence and the senses were in conflict with the method's quest for objectivity.

Until circa 1930, acousticians such as Wallace C. Sabine at Harvard University, Vivian Chrisler at the Bureau of Standards, or Harvey Fletcher of Bell Laboratories determined by listening whether a sound was louder, equal, or less audible than a given impulse, a method that was "laborious, very tiresome for the observer, and depended upon the observer's ability always to stop the watch just as the sound disappeared." 30 As of 1928, once the impulses of the tapping machines hammering onto floors had been standardized, the scientists resented even more the fact that individual human hearing was still part of the measuring process. Sound was under scrutiny to the extent that its measurement could no longer depend on human sensation. Soon afterwards, sound measurements came to be taken automatically, and the physicist's listening presence was no longer required. But, as I will show, human hearing did reappear in acoustic testing.

f.5 Lothar Cremer's "Vergleichshammerwerk," a comparative device for the simple on-site assessment of sound transmission (1957).

30 V. L. Chrisler, "Acoustical Work of the National Bureau of Standards," *Journal of the Acoustical Society of America*, 7, no. 2 (1935), pp. 79–87; here p. 83.

In the postwar period, with the proliferation of soundproofing technologies and regulating standards, a faster, easier, and simpler method for testing wall and floor constructions was called for. Outside the laboratory, on-site, things needed to be practicable. Lothar Cremer, Europe's most renowned twentieth-century acoustician, repeatedly demonstrated an affinity for, if not insistence on, listening practices, as did many of his fellow acousticians—despite the fact that the human ear had been excluded from objective scientific methods. Listening came in

handy when engineers were looking for a simpler method. Cremer, who is best known among architects for his collaboration with Hans Scharoun on the Berliner Philharmonie in 1956, collaborated that same year with the German housing secretary Bernhard Wedler on a campaign for better sound insulation in residential buildings. Wedler had formerly been responsible for construction regulations, including the DIN 4110 of 1938, for testing new construction methods, and the DIN 4109 of 1944, which dealt solely with sound insulation in buildings, and he stands out as an engineer and administrator who engaged deeply with the acoustic qualities of buildings. 31

In 1957, the Ministry of Housing printed 100,000 copies of his brochure *Baut ruhige Wohnungen* (Build Quiet Apartments). 32 Around the same time, he produced an educational film, for which Cremer provided the script. 33 The 1957 brochure features a "simple apparatus for estimating ... impact sound insulation," which Cremer had devised with his collaborators, and which operated on the basis of comparing, by ear, the sound transmission from the room above with the sound emission of this simple apparatus. 34/f.5/f.6

Acoustic standardization brings with it larger questions: Can a technical standard implement a social regime of noise control—or is it rather that the social regime legitimates the rule, which in itself has no normative power? In the case of technical standards, any normative level is arbitrary and thus artificial, as French philosopher of science Georges Canguilhem argued. 35 And yet, the issue of soundproofing and privacy was fought over so vehemently during the twentieth century that one might arguably consider acoustic standards to be valid indicators of social norms as well as agents in the process of defining them. 36 When is a noise

f.6 Information brochure for architects, issued by the German Ministry of Housing (1957).

31 DIN 4109 ("Richtlinien für den Schallschutz im Hochbau") by the Deutscher Normenausschuss (1944, single sheets reprinted in January 1959 on the occasion of a revision, annotated in December 1960). From the private archive of Dr. Joerg Wildoer, Genest AG, Berlin.

32 Bernhard Wedler, *Baut ruhige Wohnungen* (Bad Godesberg: Bundesminister für Wohnungsbau, 1957).

33 Lothar Cremer, *Schallschutz im Wohnungsbau* (Munich: Institut für Film und Bild in Wissenschaft und Unterricht, c. 1953), approx. 15 mins.

34 Manfred Heckl and Heinz Westphal, "Einfaches Gerät zur Abschätzung des Trittschallschutzes (Vergleichshammerwerk)," *Bundesbaublatt*, 9 (1957), pp. 458–61. An exemplar of such a comparative tapping machine ("Vergleichshammerwerk") was standing in a corner at the acoustic testing facility of the Technical University Berlin when I visited there in 2009, without anyone being able to tell me what it was for. The filing record was stamped: "To be discarded." Nevertheless, the newly identified machine aroused the interest of the engineers who today are equipped with maximum precision instruments and yet still rely on what their own ears hear. A restaging of the test situation at TU Berlin in June 2015 confirmed the practicality and yet the arbitrariness of such simple auditory assessment by the human ear.

35 Canguilhem elaborated the notion of "vital" and "social" norms versus "technical" standards: Georges Canguilhem, *The Normal and the Pathological*. Trans. Carolyn R. Fawcett (New York: Zone Books, 1991 [1966]), pp. 237–56. See Henning Schmidgen, "Über Maschinen und Organismen bei Canguilhem," in Georges Canguilhem, *Wissenschaft, Technik, Leben: Beiträge zur historischen Epistemologie* (Berlin: Merve, 2006), pp. 157–78; here p. 171.

36 See Sabine von Fischer, "Dynamique ou uniformisation?" (see note 24).

too loud? Is there a democratic norm for loudness, or does the right to noise stand for privilege and power, whereas the powerless are subject to a rule of silence? [37]

37 See Mark M. Smith, *How Race is Made: Slavery, Segregation, and the Senses* (Chapel Hill: University of North Carolina Press, 2006).

Wedler's brochure of 1957, as well as further literature related to building regulations, promoted the "Vergleichshammerwerk" (comparative tapping machine), despite the fact that objectivity was assumed in the 1950s to be a prerequisite of any technical science or engineering operation. This manually operated tapping machine used to assess sound insulation with the help of the human ear, by hearing the difference in loudness between two rooms, weighed 12 kilograms and cost 380 DM. [38] Together with the "Norm-Hammerwerk" (standard tapping machine), weighing 15 kilograms and costing 450 DM, impact sound measurements could be taken on site without the use of heavier and more expensive electronic equipment. Including the ear was justified by the practicality of easier handling and simple and quick measurements even by laymen. At the nexus of the technical and the social, a testing apparatus dependent on human hearing seemed acceptable to both technicians and the building administration and was therefore introduced into the otherwise mechanical, automated, and standardized practice of architectural acoustics.

38 Bernhard Wedler, *Berechnungsgrundlagen für Bauten*, 23rd ed. (Berlin: Ernst, 1959), p. 440.

Cremer's "comparative tapping machine" confronts us with the paradox of putting mechanical expertise at the service of everyday functionality and comfort in domestic and working environments. It illustrates how the inclusion of a perceiving organ in the practice of sound measurement resulted in the latter's exclusion from science's normative standards. Cremer's device is not mentioned again in the literature of the following decades, nor does anyone recall an event of its use. [39] It remains a forgotten episode, in which the scientific methods of architecture were tested for their capacity to include the listening ear as part of the apparatus for assessing a building's functionality. The use of tapping machines to test a building's acoustic performance replaced the everyday nuisance of bothersome noises by an objectively quantifiable level of loudness, which then could be measured against a technical standard. With the advent of automation and standardization, the hearing ear was excluded from acoustic testing. These endeavors in scientific enquiry were undertaken in behalf of bodily health and comfort; and yet, episodes of reintroducing hearing into the evaluation of sound, such as Cremer's "Vergleichshammerwerk," remained of little consequence and were soon disremembered. This leaves us with the question as to whether the tapping machine as an architectural device is related to the presence, or to the absence, of experience.

39 As revealed by questions I put to the older collaborators at the acoustics department of the Technical University of Berlin in June 2009 and June 2015, as well as by the collections in German museum archives.

S — 399390: Challenging the Museum with Sarah Oppenheimer
Jacqueline Maurer

In the Grand Hall of the Musée d'Art Moderne (Mudam) on the Kirchberg in Luxemburg: depending on the weather and the time of day, grids composed of trapezoids, parallelograms, and rectangles cast shadows of varying intensity on the honey-colored limestone walls and floors. It was in this location roofed by a geometrically latticed glass pyramid, and in light of the wider exhibition context, that the US-American artist Sarah Oppenheimer staged an operation in winter and spring 2016. In this work Oppenheimer departed from the motif she had pursued since 2002 in Typology of Holes: her piercing of museum walls with openings, a procedure based on the contingencies of the spatial matrix or, to use the artist's own term for these, "the array." Instead, at the invitation of Mudam, two walkable and movable spatial elements — so-called "switches" — were created for the Grand Hall. As with her previous work, this new category of spatial manipulation bears an alphanumeric title: "S" stands for "switch" while the subsequent series of digits is generated from the type and composition both of the space housing the exhibit and the rooms adjacent to it.

S-399390 is based on two primary grid systems in Mudam: the building structure and the planimetric organization of the building. The ground plan of the expansive museum, opened in 2006 and designed by the Chinese-American architect I. M. Pei, engages with the arrowhead plan of the eighteenth-century Fort Thüngen, thereby responding abstractly to the historical context of the site. The spatial axes already shift repeatedly, by 45 or 90 degrees, before visitors reach the main hall via a ramp and the reception area. Mudam's Grand Hall — the central point of orientation in the museum, and simultaneously an imposing space in which to linger and to exhibit — has been rotated 45 degrees from the primary axis of procession. Visitors are accordingly able to enter and exit the Grand Hall at three of its four corners. Coming from the reception area, the first entrance to the Grand Hall is audaciously mediated: a first-floor balcony projects diagonally into the 43-meter-high hall, delaying experience of the latter's impressive height and the rotation of its axis. This is a moment of transition or a threshold. The balcony is skewed 45 degrees above the hall's square-shaped ground plan. Extending over two stories the walls are clad with Magny Doré limestone, pierced with crystalline, triangular, or vertical skylight openings, the last of these dissolving entire walls. The Grand Hall is topped by a towering, stepped glass pyramid that echoes — like the limestone — Pei's design for the

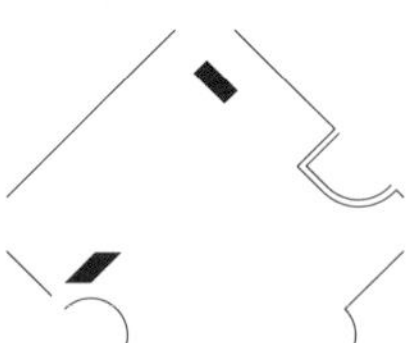

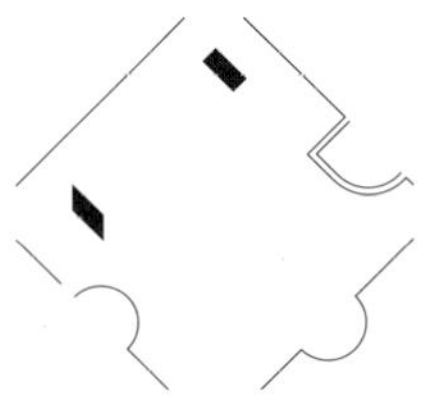

Louvre, but is crowned in this case by a glazed lantern. The fenestration connects the interior and the exterior, establishing precisely selected references to the architectural and landscape surroundings while simultaneously exposing the space to the shifting natural light of the sky.

In February 2016 Oppenheimer placed within this given architectural constellation two glazed walkways as spatial thresholds that formally correspond to the order of Pei's building. One of these "switches" has a transverse-rectangular base and a parallelogram as its counterpart ceiling profile; its companion's elements are arranged in exact reverse. Mirroring the dark coloration of the glazing framework, the dimensions of the profiles of the floor and ceiling panels are reminiscent of H-beams. Nevertheless, an impression of lightness is created, deriving from the large lateral glass walls, whereby one from each consists of a transverse rectangle and its opposite of a parallelogram.

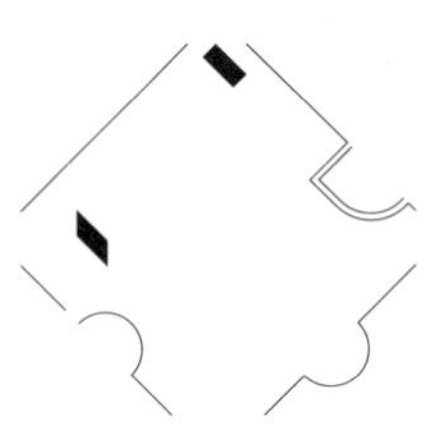

By means of the oblique walls on one of their sides the "switches" take on a rotational impulse and a directional momentum. They invoke the various grid arrangements both inherent to and produced by the architecture of the Grand Hall while simultaneously challenging them. Furthermore, the "switches" react to the existing and the newly staged threshold areas. There is a visual ambiguity about where the rooms extending from the hall, the steps to the upper story, and the spiral staircase to the basement actually lead. For the duration of the exhibition, the "switches" were relocated twice a week within the transitional verges of these spaces—or, in their own terminology, were "switched over"—and they thereby intervened in the existing threshold situations, highlighting, redirecting, transforming, and multiplying them.

Oppenheimer always aims to heighten the spatiotemporal perception of the spaces in which her interventions are staged. To this end, she investigates the way the existing spaces are visited and used by different publics, and the way in which the location influences the movement both of visitors and internal and external employees. In S-399390 the artist explores how the positions of the "switches"—which themselves contain a threshold effect—affect the succession of thresholds in the museum, as well as the direct routes that visitors so often seek to take. The precise design, principally consisting of glass, aluminum, and plywood, coupled with these varying positions, is the result of an elaborate preparation process that is a general hallmark of the artist's work. The system involved is by no means simply a negotiation of the formal architectural circumstances; rather, it consists to a far greater extent in tracing the unnoticed codifications of the respective exhibition locations in order to make them perceptible. Oppenheimer

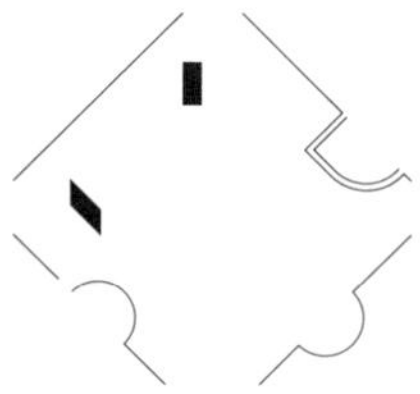

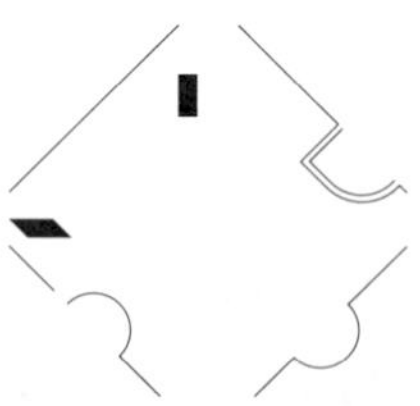

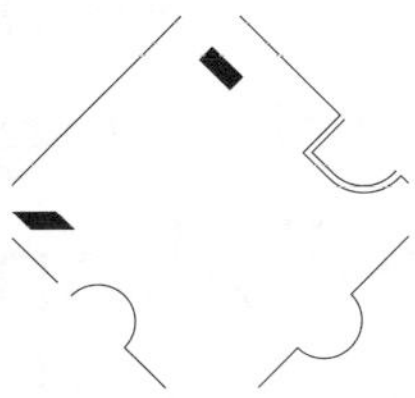

examines the various prevailing circumstances: the architectural conventions, the industrial and material norms, as well as the ideological implications to which exhibition establishments are subject. The groundwork includes studying the museum plans, researching visitor behavior on site, and above all conducting exchanges with the different staff groups, in a process of communication about utilization.

Sarah Oppenheimer understands architecture explicitly as a machine. Her oeuvre draws attention to the performativity of exhibition locations, which per se determine a very specific and highly disciplining use, and thus harbor a latent ideological statement. The "existing architecture" as formed and controlled by the visitors is factored in while the visitors themselves are activated by the artistic operation, induced to see. The overall spatial impression was constantly

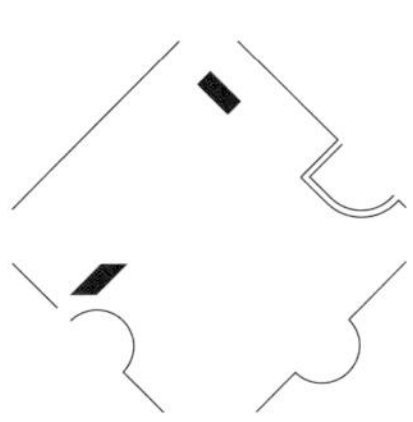

transformed by the reflections of Mudam's architecture and the visitors in the shifting seasonal daylight on the at times transparent and at times opaque appearance of the successively layered glass surfaces. Amplified by their calculated repositioning, the elaborately conceived "switches" perpetually generated and mediated new spatial configurations.

An element in Oppenheimer's conceptual logic is that the work can be relocated and recalibrated according to varying spatial scales. Because the ground plans of all the exhibition spaces in Mudam

have the same order, "the array" remains in itself the same, enabling the work to be modified. The "switches" could be dismantled after the exhibition and can be reassembled for a new presentation with new parameters.

f.1 Sarah Oppenheimer, S-399390, 2016. Glass, metal, wood, and existing architecture. Variable dimensions.
Art intervention at the Grand Hall of Mudam Luxemburg (February 2 to May 29, 2016). Photography by Serge Hasenboehler.
f.2 Changing positions of the "switches" during the exhibition at Mudam. Diagrams by Sarah Oppenheimer.

"The Mechanization of Adornment" Revisited; or, The History of the Ornament-Machine

Spyros Papapetros

What if machines not only supported the core and main programmatic functions of architecture, but also expanded to its outer surface layers, which have no clearly definable programmatic goal? Could we then gradually acquire a different view of the machine's "command" of architecture, one that while subsisting on peripheral ornamental details would nonetheless allow a glimpse of the main interior mechanisms that have propelled the precipitous "building up" of mechanization in the architecture of the last two centuries? This research moves not only from mechanical structure to machine-made ornament, but also from machine to mechanization, that is, from object to process, including historical, evolutionary, and psychological processes. It proposes an interior view of mechanization via the historiographic description of adornment's evolutionary trajectory in modernity, from organ or implement to non-functional vestige. Moreover, it contributes to a history of mechanization's impact on human life via the interface of adornment and its uncanny proximity to the body—a body that strives to amplify its power via additional corporeal enhancements, yet ultimately becomes paralyzed by their elaborate mechanical contraptions.

Spyros Papapetros is Associate Professor of History and Theory at the School of Architecture at Princeton University.

In the "Mechanization of Adornment" chapter of his magnum opus *Mechanization Takes Command,* Sigfried Giedion describes a crucial moment in the mid-nineteenth century when the machine steps beyond the public realm of factories and spaces of collective production and invades man's most "intimate" sphere: the private layer of "adornment" extending from personal furnishings and small domestic artifacts to ornamental patterns on the human skin. [1] Even though the majority of the artifacts Giedion describes in detail would be normatively labeled as decorative or ornamental, the historian chooses neither the term "decoration" nor "ornamentation" but rather "adornment"—in the German version of his manuscript "Ausschmückung." [2] The German term *Schmuck* normally designates objects used to adorn or embellish the body, from jewelry to tattoos, while *Ausschmückung* connotes the expansion of the same process in space. The term marks the transition from an isolated ornamental object to a decorative system that extends from the individual body to the garnishing of an architectural environment: moldings, carpets, curtains, and wallpaper, all of which project the bodily layer of adornment to the inorganic envelope of a building via the auspices of industrialized production. Where adornment originally functioned as a portable ecological system, allowing human

1 Sigfried Giedion, "The Mechanization of Adornment," in Giedion, *Mechanization Takes Command: A Contribution to Anonymous History* (New York: W. W. Norton & Company, 1969 [1948]), pp. 344–63.

2 Sigfried Giedion, "Mechanisierung der Ausschmückung," in Giedion, *Die Herrschaft der Mechanisierung: Ein Beitrag zur anonymen Geschichte* (Frankfurt: Europäische Verlagsanstalt, 1982), pp. 381–402.

subjects to collect and rearrange pieces of their environment into a meaningful cosmic order and then apply this to the human body, [3] the mechanized adornment of the nineteenth century as described by Giedion expresses the repressed ecological potential of modern technological environments whose efforts to mediate between mechanized production and the private sphere become lost under the cover of meaningless ornamental layers produced assiduously by the machine.

3 See my description of ornament as "portable ecology" in "An Ornamented Inventory of Microcosmic Shifts: Notes on Hans Hildebrandt's Book Project 'Der Schmuck' (1936–37)," *The Getty Research Journal*, 1 (2009), pp. 87–106; here p. 96.

For Giedion, the closer the machine gets to the body, the more its architectural "symptoms" become pathological. Giedion describes an era which saw a shift from the pathological origins of the machine as an inorganic prosthesis propping the ailing human body to a state in which the machine itself becomes pathologized, and keeps on reproducing its symptoms or even spreading its viruses to the human body. As a cultural historian, Giedion diagnoses not only a form of mimicry between adornment and the body but also a pathological, analogizing intimacy between adornment and the machine.

For Giedion, adornment is a "wish" that is "innate" and "ineradicable;" something mechanical and instinctive, like "hunger" or "love" (meaning apparently love-making or sex). [4] The machine capitalizes on such instinctive impulses to feed or even overstuff the body with an abundance of previously luxurious articles that have now become widely accessible via the auspices of industrial production. This process has led to both a proliferation and a devaluation of ornament, which now retains only a vestige of its former economic prestige. Giedion seems to suggest that it is precisely the accumulation of these mechanically multiplying layers of adornment that will eventually bury architectural ornamentation and lead to the unornamented cladding of building surfaces by early twentieth-century architects after Adolf Loos; yet, prior to this phenomenal eclipse of ornament, a number of intermediate transitions will occur, which, for the historian, serve as collective psychological symptoms. "Here as scarcely anywhere else are displayed the fears and preoccupations of a period," writes Giedion, as if the curvilinear forms of the fanciful artifacts produced en masse in the late nineteenth century were psychograms of that era's anxious *Weltanschauung*. [5]

4 Giedion, *Mechanization* (see note 1), p. 344.

5 Ibid., p. 349.

If, through its exhaustive examination of bathtubs, elevators, and poultry de-feathering machines, Giedion's *Mechanization Takes Command* presents an "antechamber" of modern architecture, then this brief, 20-page section on the mechanization of adornment might in fact tell a slightly different story. The chapter on adornment discloses not only what comes before but also what might follow modern architecture, which had erupted

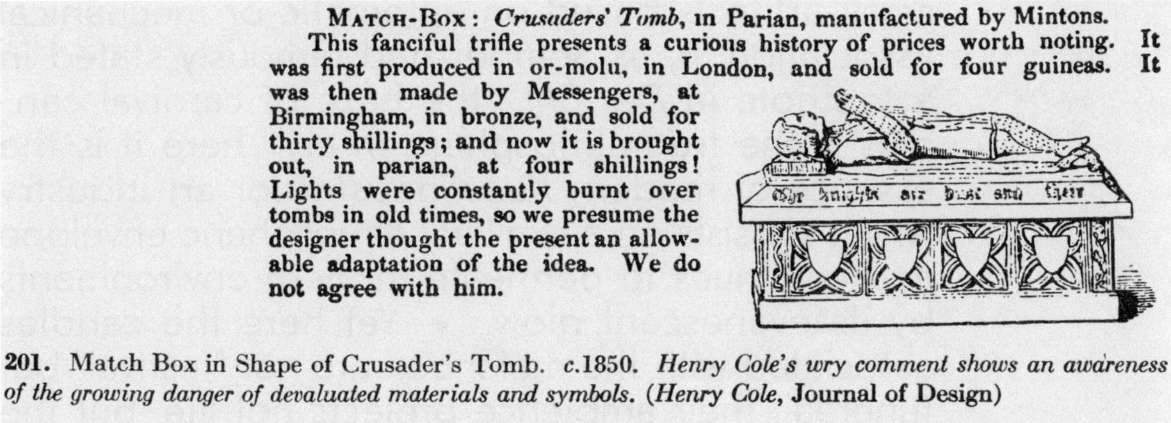

MATCH-BOX: *Crusaders' Tomb*, in Parian, manufactured by Mintons. This fanciful trifle presents a curious history of prices worth noting. It was first produced in or-molu, in London, and sold for four guineas. It was then made by Messengers, at Birmingham, in bronze, and sold for thirty shillings; and now it is brought out, in parian, at four shillings! Lights were constantly burnt over tombs in old times, so we presume the designer thought the present an allowable adaptation of the idea. We do not agree with him.

201. Match Box in Shape of Crusader's Tomb. *c.*1850. *Henry Cole's wry comment shows an awareness of the growing danger of devaluated materials and symbols.* (*Henry Cole,* Journal of Design)

f.1 "Match Box in Shape of Crusader's Tomb. c. 1850," *Journal of Design*, edited by Henry Cole; reprinted in Sigfried Giedion, *Mechanization Takes Command*, 1948.

30 years earlier but by the mid-1940s was already losing steam and entering (somewhat prematurely) its own pathological stage. And, if modern architecture was in fact heralded by an ostensible ban on ornament, imposed by Loos, it was precisely the return of ornament—initially manifest in scholarly attention to the designers of Art Nouveau, then shortly afterwards in the building practice of postwar designers—that would celebrate the apparent end (or death) of modernism by a decorated procession of highly ornamented edifices. Here, one might start deciphering correspondences between Giedion's chapters on "The Mechanization of Death" (on industrialized slaughterhouses and hog execution machines) and the "Mechanization of Adornment," as both accounts demarcate an ending that coincides with a festively decorated revival. For Giedion, mechanized adornment functions as an ambivalent symptom. From statuary to pottery, from carpets to wallpapers, this mechanized architectural envelope oscillates between life and death, flatness and depth, the virtual and the real, and the plastic and the graphic, including the typographic. The same ornamental artifacts can be endlessly reproduced in silver or tin, "marble or plaster; china or papier mâché" [6] in a perpetual, yet incongruous *Stoffwechsel*—the term used by Gottfried Semper to connote material change facilitated by the preservation of a formal pattern.

6 Ibid., p. 345.

Giedion cites the example of a small metal matchbox modeled after a Gothic "crusader's tomb" with a recumbent *gisant* sculpture of the deceased crusader sculpted on its top. Giedion discovered a drawing of this artifact at the bottom of a page of the first volume of Henry Cole's *Journal of Design and Manufactures* published between 1849 and 1852—one of the main textual sources for Giedion's study of nineteenth-century ornamentation. [7/f.1] The whimsical analogy between a matchbox and a crusader's tomb is prompted, according to Cole, by the image of "lights [that] were constantly burnt over tombs in older times." [8] This quasi-surrealist analogy demonstrates that, in spite of all its fluctuations, there is a consistent logic behind the industrial reproduction of these

7 In a footnote Giedion mentions that he had dealt more extensively with Cole and "the reforming movement of 1850" in an "unpublished study *Industrialisierung und Gefühl* (Industrialization and Feeling), based on the manuscripts and diaries of Cole preserved in the Victoria and Albert Museum, London." Ibid., p. 347n.

8 See caption by Cole reproduced in ibid.

small artifacts based on automatic or mechanical association. If, as Semper had famously stated in a footnote in *Der Stil*, "the haze of carnival candles is the true atmosphere in art," here it is the essence of modern *Kunstindustrie* or art industry which subsists on a similarly atmospheric envelope and continues to permeate modern environments by its evanescent glow. [9] Yet here the candles associated with the matchbox are not "festival" but funereal; their ambience projects not life, but the glimpse of an afterlife, or eternal life, following the death and transubstantiation of the organic body into an industrially produced unit: the matches lying flat inside the matchbox.

Not only flickers of bygone artifacts but also the historical figures of architects and designers such as Cole, Semper, and Owen Jones suffuse Giedion's chapter on adornment. And yet while the historian devotes several pages to Cole, Jones, and the 1851 Great Exhibition, he writes no more than a page on Semper, who had likewise participated in the Great Exhibition and was renowned for his theorization of ornament and his copious analysis of small artifacts, both as material and cultural signifiers. Though Giedion praises Semper as the "sole" designer who "attempted to mold the experience and principles ... of the Great Exhibition into a broad system," he criticizes the "utilitarianism" that "colored his interpretation of historical epochs." [10] Giedion also argues that Semper "was gifted enough to systematize the intellectual views of the [eighteen] 'fifties," yet he questions whether "these attitudes were comprehensive enough to be forged into a system." The reason for such failure is that "[a]n insurmountable barrier of feeling kept [Semper's] generation from perceiving the pure forms latent in machine-made objects." [11]

It may be true that Semper was ambivalent about the effects of the machine on the making of artifacts, particularly in his review of the Great Exhibition in his publication, "Science, Industry, and Art;" [12] yet in his *Style* he refers to a "special style" that "needs to be created for the machine" while cautioning nonetheless that "the critical factor [for such style] should be function, measured independently of the machine." [13] But it was also Semper who,

9 Gottfried Semper, *Style in the Technical and Tectonic Arts; or, Practical Aesthetics*. Trans. Harry Francis Mallgrave and Michael Robinson (Los Angeles: The Getty Research Institute, 2004), p. 439. On the same phrase see Caroline van Eck, "Figuration, tectonics and animism in Semper's *Der Stil*," *The Journal of Architecture*, 14, no. 3 (2009), pp. 325–37.

10 Giedion, *Mechanization* (see note 1), pp. 357–8.

11 Ibid., p. 359.

12 Gottfried Semper, "Science, Industry, and Art," in Semper, *The Four Elements of Architecture and Other Writings*. Trans. Harris Francis Mallgrave and Wolfgang Herrmann (Cambridge: Cambridge University Press, 1989), pp. 130–67.

13 Semper, *Style* (see note 9), p. 654.

following a stint as a teacher at London's School of Design, and in interaction with various British theorists of ornament, [14] consequently theorized ornamentation in his 1856 lecture on the symbolism of adornment—an essay that Giedion does not mention. [15] Based on the laws of the sciences of Dynamics and Statics describing the impact of natural forces such as Gravity (*Schwerkraft*) on the form of natural bodies, Semper devised in his essay a series of regular principles that should apply to all artifacts of human ornamentation.

Once again the form of an artifact is prescribed by mechanical forces, but here such forces follow the mechanics of nature, which do not always synchronize with those of human invention. Semper's ultimate objective was to restore a form of alignment between nature and culture, but unless one subscribes to Comtean positivism, the laws of mathematics and physics can never be perfectly aligned with those of human economy, politics, and culture. It is precisely the strife and divergence between these conflicting cosmic faculties that become visible in the form or even the texture of decorative artifacts. Like Semper's "directional ornaments" (*Richtungsschmuck*), adornments essentially function as projective devices, indexes of the bodily movements of their human carriers as well as the larger mobility of civilizational forces. [16] In effect, they are far more sensitive to change than built monuments, whose typological arrangements, while expressive of sociopolitical hierarchy, require greater time to register such cultural transformations. The decorative objects of primeval human industry are treated by Semper not simply as an aggregate of uncoordinated individual artifacts, but as a thoughtfully articulated spatial system whose directional principles will later be retraced in those directional elements of architectural structures, such as corridors, atria, and walled entrances. [17] To describe the precedence of objects as cultural signifiers *before* architecture, Semper invented in *Style* the term "prearchitectonic conditions" (*vorarchitektonische Zustände*) applied to origins, technique, or an entire period of time, describing a state of culture in its "earliest stages" when artifacts, such as weapons, implements such as pottery, and adornments are present, but massive architectural monuments, such as those known from Egypt and Assyria, are not. [18] Such a "prearchitectonic" state describing social relations mediated by systems of objects is by no means the opposite of architecture, but rather its evolutionary prelude, tracing a set of possibilities in which some may come into being while others perish, only to return in a modified form.

In fact, Giedion's project in *Mechanization* appears to be parallel to that of Semper; like the nineteenth-century architect,

14 Elena Chestnova, "'Ornamental design is … a kind of practical science': Theories of ornament at the London School of Design and Department of Science and Art," *Journal of Art Historiography*, 11 (2014); https://arthistoriography.files.wordpress.com/2014/11/chestnova.pdf (accessed February 29, 2016).

15 Gottfried Semper, "Über die formelle Gesetzmässigkeit des Schmuckes und dessen Bedeutung als Kunstsymbol," *Monatsschrift des Wissenschaftlichen Vereins in Zürich*, 1 (1856), pp. 101–30; republished in Semper, *Kleine Schriften*, ed. Hans and Manfred Semper (Berlin: W. Spemann, 1884), pp. 304–43. The first section of the essay has been translated into English by David Britt as "From Concerning the Formal Principles of Ornament and Its Significance as Artistic Symbol," in Isabelle Frank (ed.), *The Theory of Decorative Art: An Anthology of European and American Writings, 1750–1940* (New Haven, CT: Yale University Press, 2000), pp. 91–104. The second section has been translated by Kathryn Shoefert and Spyros Papapetros as "On the Formal Principles of Adornment and its Meaning as a Symbol in Art (second section)," in *RES: Anthropology and Aesthetics*, 57/58 (2010), pp. 299–308. For a commentary on the legacy of Semper's essay, see my "World Ornament: The Legacy of Gottfried Semper's 1856 Lecture on Adornment," ibid., pp. 309–29.

16 Semper, "Über die formelle Gesetzmässigkeit des Schmuckes" (see note 15), p. 113.

17 See the late lecture "On Architectural Styles," in Semper, *The Four Elements of Architecture* (see note 12), pp. 264–84.

18 Gottfried Semper, *Der Stil in den technischen und tektonischen Künsten, oder, praktische Aesthetik*, vol. 1 (Frankfurt: Verlag für Kunst und Wissenschaft, 1860), p. 229.

the twentieth-century architectural historian describes modernism's "prearchitectonic conditions"—a set of formative and tectonic principles based on the history of industrially produced objects which will later find expression in the spatial arrangement of modernist buildings. In spite of Giedion's and other modernist architectural historians' mistrust of the stylistic terminology associated with architectural practices and art historical methods of the nineteenth century, we could even describe *Mechanization Takes Command* as the *Style* of the twentieth century—a voluminous contribution to our knowledge of "the technical and practical arts" from a bygone era serving as models for the architecture that is about to follow.

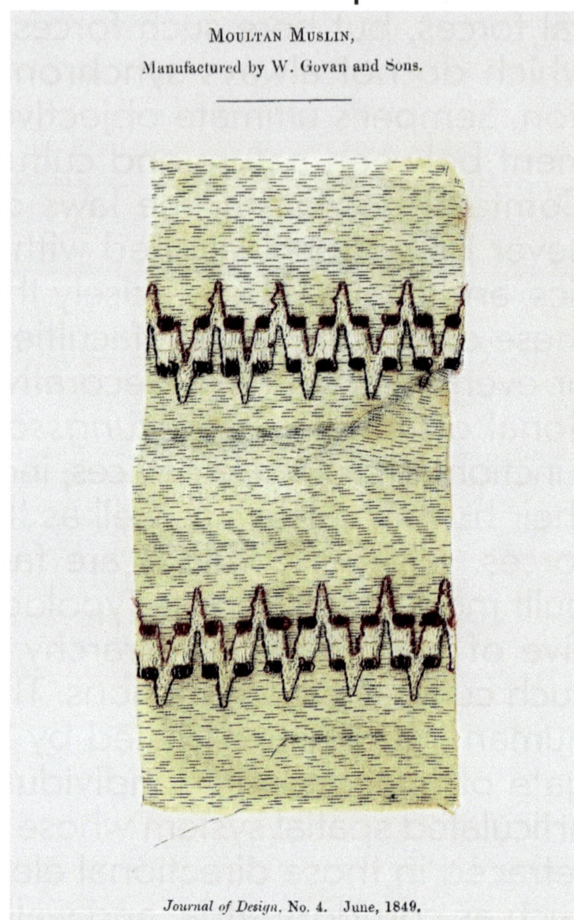

f.2 "Moultan Muslin, Manufactured by W. Govan and Sons"; fabric sample attached to *Journal of Design*, June 1849.

"It began with the carpets" writes Giedion, once again echoing Semper in his text on the mechanization of adornment. 19 A new type of machine loop allowed the indiscriminate transference of ornamental patterns among the most divergent materials and scales, as for example, the industrial production of carpets with enormous roses or other flowers, several times the size of the original natural specimen. With the advent of the machine, these naturalistic (yet essentially unnatural) floral patterns started to indiscriminately proliferate, from book binding covers to wallpapers and, finally, rugs, the patents for all of which appear in Cole's *Journal of Design and Manufactures*, not as printed illustrations but as physical samples made of colorful fabric or paper and attached directly to a page. f.2

19 Giedion, *Mechanization* (see note 1), p. 345.

Giedion also notes that in spite of their liberal placement on the surface of the carpet (eschewing the formal symmetrical arrangement of carpet design in earlier periods), floral patterns are still used as spatial markers for the placement of domestic furniture. Even if flat, the flower motifs become space creators by articulating decorative correspondences among all elements of the decorated interior, whose surfaces start reverberating with one another. The vegetal patterns' perpetual rotation from the horizontal to the vertical plane does not enhance gravity, but rather defies it, contrary to Semper's lawful principles of adornment.

From "prearchitectonic conditions," we abruptly transition to a post-architectural state where weight, scale, and material matter all too little, and where carpets, furniture, and flowers appear to float within that which Walter Benjamin, in his references to Jugendstil dream interiors, described (in the words of Paul Morand) as an "aquarium" or a "submarine" environment. [20]

Ornament's plastic expansion in space is facilitated by printed media, such as books, folios, and most prominently, journals, using the latest advances in typography and drawing reproduction. Apropos Owen Jones's popular *The Grammar of Ornament*, Giedion notes that the British designer "carefully assemble[s] ornament from widely separated peoples and periods" then "lifts the ornament from its material ... and flattens it out in color upon the plane surface." [21] Yet it is precisely the flattening of ornament on the printed page that further facilitates its reproduction in other mediatic surfaces such as carpets, upholstery, and wallpaper, which start mirroring as in a *mise en abyme* the ornamented patterns and layout of the lavishly decorated journal. Such phenomenal explosion of ornamentation via graphic media in the second half of the nineteenth century also portends its imminent eclipse. Before Loos would seek to ban ornament in 1908, the enormous piles of publications on ornament had already buried it. Such media ostentatiously display ornament on their pages but by the same token gradually occlude its presence in the real world. The rigorous classification and compartmentalization of ornament from around the world in publications such as Owen Jones's *Grammar* underlines that ornament gradually becomes the object of museum collections with little relation to the modern way of life. [22]

Giedion brilliantly captures the dialectics between the mechanization and gradual rudimentation of ornament in his investigation of the origins of adornment's printed culture. As mentioned earlier, Giedion's main bibliographic source for his chapter on mechanized adornment is Cole's *Journal of Design and Manufactures*, one of the first periodicals exclusively devoted to decorative design in the applied arts. Cole's journal ran for only six issues, and yet its brief life was protracted by a series of afterlives and reincarnations in other similar journals. Giedion hails *The Journal of Design* as "the *Esprit nouveau* of 1850," referring to Corbusier's and Ozenfant's modernist journal published 70 years later. [23] Cole's journal is filled with the decorative artifacts and ornamental motifs that Corbusier would castigate or caricature in the pages of *L'Esprit Nouveau* and his book *L'Art décoratif d'aujourd'hui* [24] —criticism whose very irony reveals the architect's ambivalence towards his own apprenticeship

20 Paul Morand, *1900* (1931) quoted in Walter Benjamin, *The Arcades Project*. Trans. Howard Eiland and Kevin McLaughlin (Cambridge, MA: Harvard University Press, 1999), p. 549.

21 Giedion, *Mechanization* (see note 1), p. 354.

22 I make this argument in my "World Ornament" (see note 15).

23 "It is the *Esprit nouveau* of 1850, except that it would embrace the whole of industry, the entire world, and was not confined like its counterpart of 1920 to a small magazine of the *avant-garde*." Giedion, *Mechanization* (see note 1), p. 349.

24 Le Corbusier, *L'art décoratif d'aujourd'hui* (Paris: Crès, 1925).

in the decorative arts while following in the profession of his father as a young watch and jewelry designer prior to becoming an architect. 25

202. HENRY COLE: Society of Arts Competition, 1845. Tea Service 'For Common Use.' *This popular tea service represents one of the early endeavors to improve public taste: 'A model of plain and cheap earthen ware. The aim has been to obtain as much beauty and ornament as is commensurate with cheapness. Ornaments on the handles are so designed as not to interfere with the simplicity of the outlines. The cup, being much deeper than wide, offers least scope for the radiation of heat.'* — Fifty Years of Public Life. (*Courtesy the Society of Arts, London*)

The gradual elimination of ornament is in fact illustrated by a semi-unornamented tea service "for common use," designed by Cole in 1845 and a drawing of which is reproduced in *Mechanization.* f.3 Named after the tea plant *Camellia*, copies of Cole's tea service "met with tremendous popularity," Giedion notes, and are "manufactured to this day." 26 The pots, cups, and saucers of Cole's tea service resemble a proto-purist design. "Ornaments on the handles are so designed as not to interfere with the simplicity of the outlines," notes Cole in Giedion's caption. 27 Indeed, ornamental decoration is limited to certain parts of the artifact, such as the spout and handles. These ornamented limbs seem almost detachable from the rest of the artifact's body, as if they were prosthetic additions whose removal would leave the rest of the surfaces plain. Here ornament becomes precisely *ad-orn[a]ment*: an additional layer such as a jewel, a piece of headgear, or a breastplate, worn on top of the organism's skin. Such adornment enhances the qualities of the industrially produced surface with its perfectly smooth and highly polished texture. Cole's tea service manifests a division of both function and labor by distinguishing *Schmuck* from *Gerät*, ornament from implement—a distinction that was to have grave consequences for the regressive development and final demise of ornament in modernist practice. Following an evolutionary logic, the non-functional ornamental appendages would gradually atrophy and vanish from modernist buildings

204. OZENFANT: Drawing, 1925. *The things we use daily become part and parcel of our life. Like the Cubists, the Purists of the 1920's pointed to the objects purified and standardized by long use, with plain and simple outlines leading to 'mariages de contours.'*

of Design. Henry Cole, who scarcely ever makes personal recommendations in his *Journal of Design*, calls to the attention of English manufacturers that Semper's 'knowledge both of architecture and of general decoration is profound and his taste excellent.' And that Semper is the man 'from whom our manufacturers would be likely to obtain great help.'[49] When later, in 1855, Semper was appointed professor to the newly founded Technische Hochschule in Zürich he systematized the expressions and experiences of London in his *Style in the Technical and Tectonic Arts, or a Practical Aesthetic.*[50] The first two volumes appeared in 1860 and 1863; the third, which was to have shown the bearing of art on social development, was never printed. For decades, Semper's stand influenced art theory. As late as 1910, the German reform movement in the decorative arts, which took fitness to purpose as its final criterion, regarded

[49] *Journal of Design*, vol. VI, p.113.
[50] *Der Stil in den technischen und tektonischen Künsten*, 1860–63, 2nd ed., 2 vols., München, 1878–9.

358

f.3 Henry Cole, "Society of Arts Competition, 1845. Tea Service 'For Common Use'"; reprinted in Sigfried Giedion, *Mechanization Takes Command,* 1948.

25 On Jeanneret-Le Corbusier's past in decorative design, see the classic study of H. Allen Brooks, *Le Corbusier's Formative Years: Charles-Edouard Jeanneret at La Chaux-de-Fonds* (Chicago: Chicago University Press, 1997).

26 Giedion, *Mechanization* (see note 1), p. 348.

f.4 Amédée Ozenfant, "Drawing," 1925—Henry Cole, "Drawings of Simple Objects for Child Education," *Journal of Design*, 1849: half-page double spread in Sigfried Giedion, *Mechanization Takes Command,* 1948.

27 Ibid., p. 351.

and artifacts, ceding place to what Giedion calls "elementary" or "standard objects as models." [28]

To illustrate such "elementary" object-models, Giedion uses an image comparison between a "purist" drawing by Ozenfant and an educational drawing by Cole. [29/f.4] Both images appear as quasi-photographic negatives, since the contours of all the objects are traced in white on a plain black background. To further highlight their morphological similarities, Giedion reproduces the juxtaposed images here in the exact same format as an upper half-page double spread: an ornamentally symmetrical arrangement that facilitates the images' conceptual mirroring or reversal. Cole's drawing placed by Giedion on the right is reproduced from an insert in Cole's *Journal* which was originally published in a book teaching children how to draw objects lucidly on a blackboard while eschewing any shadows or decorations. Giedion reads this plain representation as an *avant-la-lettre* "purist" approach to object-making presaging Ozenfant's method demonstrated on the previous page.

205. HENRY COLE: Drawings of Simple Objects for Child Education. *For teaching school children, Cole recommends drawing with thick chalk on a blackboard, omitting detail. Their observation will be trained by rendering serial-produced articles of daily life: bottles, jugs, glasses, etc.* (Journal of Design, *vol.* I, 1849)

him as a basic authority. To the utilitarianism that guided the English reformers, he owed his starting point: that practical art existed before architecture, as he set forth in an essay written in London in 1853. Utilitarianism colored his interpretation of historical epochs. Semper ranks among the few significant architects of his time, and he was gifted enough to systematize the intellectual views of the 'fifties. It is questionable, however, whether these attitudes were comprehensive enough to be forged into a system. An insurmountable barrier of feeling kept this generation from perceiving the pure forms latent in machine-made objects.

Only through fragmentary utterances did one glimpse the abstract forms inherent in industrial production. *The Times* article, which clearly bears the mark of Cole's circle, says of the Great Exhibition: 'Some sections, especially that of machinery, feeling their preeminence secure and undoubted, have been content to be plain and unpretending. The only beauty attempted is that which

359

In Cole's drawing, the composition consists not in the combination of heterogeneous part-objects but in the recombination of heterogeneous whole objects: a jug, a glass, a brush, a clothing iron, but also a boot, which is the only clothing item among all the service ware and other implements depicted. And yet, by being produced as mere outlines, these objects appear to jettison any content. They are decoratively arranged on the drawing's rectangular frame so that they complement one another as mere shapes, emptied of their utilitarian functions. They are implements treated as ornaments and yet they still function as pedagogical models for how an object ought to appear. The elementary lesson being taught here is not only how to draw artifacts but also how to reconfigure them into a network of ornamental relations. We retrace the movement "from ornament to object" recently

28 Ibid., p. 356.

29 Ibid., pp. 358–9.

mapped by Alina Payne as "a genealogy of architectural modernity," but in reverse: Cole's pedagogical drawing projects the surreptitious return of the object of mechanized modernism back to ornament. 30

f.5 Interior view of Le Corbusier's pavilion for *L'Esprit Nouveau* (1925), in *L'Architecture vivante,* Fall/Winter 1925.

30 Alina Payne, *From Ornament to Object: Genealogies of Architectural Modernity* (New Haven, CT: Yale University Press, 2012). See also my review of Payne's book, "Ornament and object — ornament *as* object," *Journal of Art Historiography*, 7 (2012); https://arthistoriography.files.wordpress.com/2012/12/papapetros-review.pdf (accessed Feb. 29, 2016).

Even if Ozenfant's drawing on the left precedes Cole's in Giedion's illustrations, the artwork by the modernist painter appears as an echo of the mid-nineteenth century "model" on the right. But it could also be the reverse. The very spine of Giedion's *Mechanization* acts as an axis for the historian's anachronic oscillation between the two images he treats here as pendants. Not only Cole's image but Giedion's entire layout is a model for instruction on how to draw; drawing here concerns not just objects but trans-historical connections between chronologically distant object models. Such historiographic *dissymétrie* (to use Roger Caillois's formal term) is a genuinely ornamental principle. 31

31 Roger Caillois, *La Dissymétrie* (Paris: Gallimard, 1973).

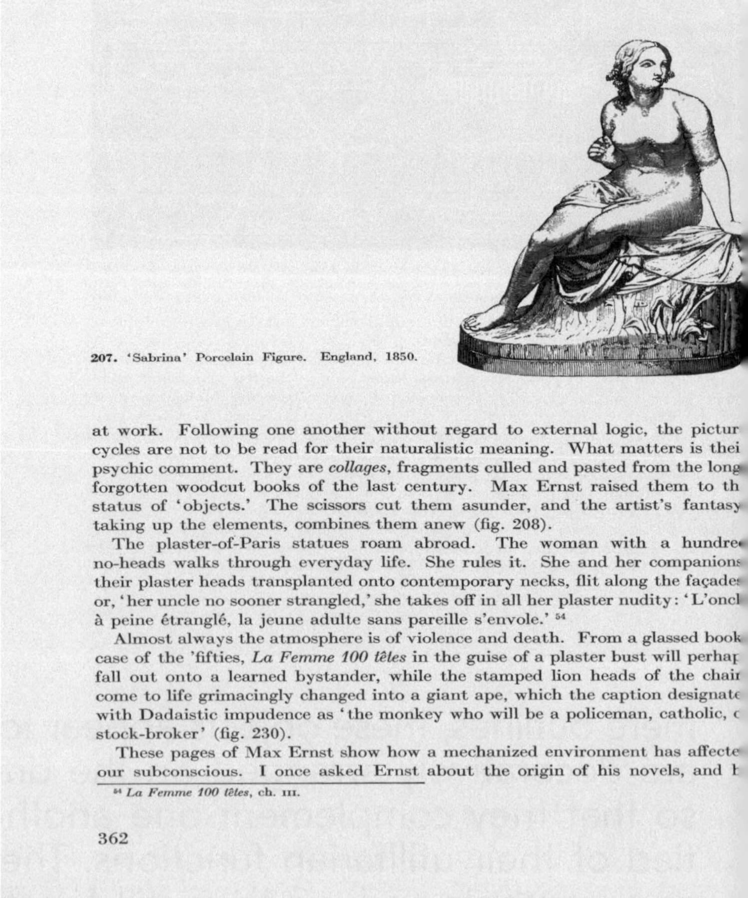

207. 'Sabrina' Porcelain Figure. England, 1850.

at work. Following one another without regard to external logic, the pictur
cycles are not to be read for their naturalistic meaning. What matters is thei
psychic comment. They are *collages*, fragments culled and pasted from the long
forgotten woodcut books of the last century. Max Ernst raised them to th
status of 'objects.' The scissors cut them asunder, and the artist's fantasy
taking up the elements, combines them anew (fig. 208).

The plaster-of-Paris statues roam abroad. The woman with a hundre
no-heads walks through everyday life. She rules it. She and her companions
their plaster heads transplanted onto contemporary necks, flit along the façade
or, 'her uncle no sooner strangled,' she takes off in all her plaster nudity: 'L'oncl
à peine étranglé, la jeune adulte sans pareille s'envole.' [54]

Almost always the atmosphere is of violence and death. From a glassed book
case of the 'fifties, *La Femme 100 têtes* in the guise of a plaster bust will perhap
fall out onto a learned bystander, while the stamped lion heads of the chai
come to life grimacingly changed into a giant ape, which the caption designate
with Dadaistic impudence as 'the monkey who will be a policeman, catholic, c
stock-broker' (fig. 230).

These pages of Max Ernst show how a mechanized environment has affecte
our subconscious. I once asked Ernst about the origin of his novels, and h

[54] *La Femme 100 têtes*, ch. III.

362

f.6 "'Sabrina' Porcelain Figure. England, 1850" — Max Ernst, "The Plaster Statues Roam Abroad. (La Femme 100 têtes, 1929):" half-page double spread in Sigfried Giedion, *Mechanization Takes Command,* 1948.

Were we to expand this ornamental reading from the textual domain of the book to an architectural space, we could look at one of Corbusier's contemporary interior spaces, such as that of *L'Esprit Nouveau* pavilion for the International Exposition in Paris of 1925. f.5 *L'Art décoratif d'aujourd'hui* or "decorative art of today" as reconfigured in the Corbusian interior subsists on a system of analogies, reflections, and correspondences between the shapes and materials of industrial artifacts, such as chairs and storage cabinets, and the shapes of two-dimensional objects and artworks, including carpets, sculptures, and paintings. While more or less stripped of conventional ornament, the interior is reproduced by the mechanical repetition of "ornamentative" connections between objects. [32] If this is indeed a model of the Corbusian "machine for living in" (*machine à habiter*), then it is so not thanks to its programmatic efficiency or the externalization of the mechanical elements of its infrastructure, but to the very afterlife of its nineteenth-century sense of correspondence, which survives in the over-ornamented ambience of symbolist literature suffusing Corbusier's writings, and also, as we see here, his architectural projects.

In Corbusier's case the relation between ornament, architecture, and machine is still an analogy, yet as Giedion describes, this relation turns into a homology when the machine starts producing ornaments and buildings that increasingly look like a machine. While, in the nineteenth century, the machine produced decorative objects that imitated handicraft artifacts which in turn imitated nature (as in the second-degree mimicry enacted in the "electro-process[ed]" leaf-shaped-candlestick mentioned elsewhere by Giedion), the twentieth-century machine unabashedly reproduced an image of itself in a series of objects. The machine itself became an ornamental "style" or, as Reyner Banham would later call it, an "aesthetic." [33] Instead of the machine (-made) ornament we are introduced to the ornament machine—an ontological reversal signaling the formal and procedural identification between adornment and the mechanical apparatus.

)8. MAX ERNST: The Plaster tatues Roam Abroad. (La emme 100 têtes, 1929)

L'oncle à peine étranglé, la jeune adulte sans pareille s'envole.

eplied: 'They are reminiscences of my first books, a resurgence of childhood memories.'

The artist fuses these childhood memories into a mirror to morality. But here are many people for whom nineteenth-century ruling taste still forms the ell of feeling, and in whose souls, as T. S. Eliot once put it, the nineteenth entury lingers on into the twentieth. Still among these are most of our conemporaries, in particular the men whose taste determines public buildings and monuments. For most of these, the origin of their esthetic faith has never come consciousness.

Do not Carlyle's lines written toward mid-century in his restive *Latter-Day amphlets* seem to echo a caption for Max Ernst?

The fact is . . . the Fine Arts, divorced entirely from truth this long while, and wedded most professedly to falsehood, fiction and such-like, are got what we must call, an sane condition: they walk abroad without keepers, nobody suspecting their bad state, nd do fantastic tricks.[55]

[55] Cole cites this passage in the *Journal of Design* (1850), vol. III, p.91.

363

Considering ornament's association with nineteenth-century theories of correspondence and the attachment of modern designers like Corbusier to symbolist poetics steeped in unconscious associations, Giedion's swift switch to surrealism in the final section of his chapter on adornment might appear less sudden; in fact it is almost automatic. Similar to the strategy he used to introduce purism, Giedion announces surrealism with a pictorial juxtaposition, this time between the engraving of a porcelain statuette of the seated figure

32 I develop a similar argument on the analogical correspondences of Corbusier's pavilion apropos Léger's painting hung inside the same interior in my "'The Most Conscientious Mason': Léger's Architectonic Analogies," in Anna Vallye (ed.), *Léger: Modern Art and the Metropolis* (Philadelphia: The Philadelphia Museum of Art, 2013), pp. 201–10.

33 Reyner Banham, "Machine Aesthetic," *Architectural Review*, 117 (1955), pp. 225–8.

f.7 Advertising brochure for Sigfried Giedion, *Mechanization Takes Command*, 1948.

of a young female named "Sabrina" from 1850 and a montage from Max Ernst's *La femme 100 têtes* (1929) titled "The Plaster Statues Roam Abroad" featuring a similar female "jeune adulte" in a fleeing pose. 34 The same figure would be reproduced on the cover of an advertising pamphlet for Giedion's *Mechanization* which bestows on this female "young adult" the status of a mechanical emblem. f.6/f.7

34 Giedion, *Mechanization* (see note 1), pp. 362–3.

Such figures, as Giedion acknowledges, leap "from a glassed book-case of the [eighteen] 'fifties," the era that produced the "first books" which artists like Ernst would have read in their childhood. 35 Surrealism for Giedion describes the uncanny moment when this entire world of vegetal and anthropomorphic ornaments spit out by the machine between 1850 and 1890 and put to sleep by the modernist aesthetic inevitably reawakens with even greater vibrancy. Of course here the ornaments return not as physical environments but as two-dimensional projections of hyper-ornamented fantastic interiors flattened on the typographic page of surrealist illustrated romans. Giedion would return to surrealism in his next chapter on "The Reign of the Upholsterer" (which is the companion chapter to his "Mechanization of Adornment") with another pictorial juxtaposition between a "surrealist interpretation of the 19th-century Interior" — taken from another montage from Ernst's *La femme 100 têtes,* in which a female figure leaps from a glass book case — and an image of Sarah Bernhardt's equally hyper-ornamented "studio," reprinted from a decoration journal of the early 1890s, in which the famous actress is seen

35 Ibid., p. 362; the phrase derives from a caption by Ernst in his *La femme 100 têtes*, quoted in ibid., p. 386.

f.8 Max Ernst, "Interior ... (La Femme 100 têtes, *Paris*, 1929)" — "Sarah Bernhardt's Studio, 1890 (*The Decorator and Furnisher*, New York, 1891)": half-page double spread in Sigfried Giedion, *Mechanization Takes Command*, 1948.

230. Surrealist Interpretation of the 19th-Century Interior, Max Ernst, 1929. *Peering from a gla bookcase, the woman with a hundred no-heads in the guise of a plaster bust startles a pensive bystander. stamped lion heads of the chairs are changed into a giant ape.* (*Max Ernst*, La Femme 100 têtes, *Paris*, 19

France's orientalizing cushion furniture, England developed a plainer sty connected with the habits of club life. These black leather easy chairs and so were destined for groups of pipe-smoking men. From this furniture, direct li lead to the reforms of William Morris and his followers and thence to the pres day. The English also did not allow the upholsterer so free a hand in th comfortable seating for the drawing room or the bedroom. But our knowled of this development is fragmentary.

For a time it seemed that the Rococo wave that gripped France in the 'thirt promised more than a stylistic revival. The chairs of the Rococo, molded to human form were elaborated in the 'sixties with an originality that cannot wholly ignored. The seats become broader, lower, and deeper. Various hyb forms arise, designed for what is neither sitting nor lying. They have one po in common: the invitation to informal posture. We shall see how truly this n static, changing posture corresponds to the character of the century.

Describing this furniture, a writer in whom the *Ancien Régime* still lives as in so many Frenchmen, cannot repress the comment as he surveys the vari species of *confortables*, 'On entering our *salons* these days (1878) one invol tarily asks oneself whether these women, these men, so *nonchalantly* reclini and sprawled, can be the descendants of that French société once outstandi for its brilliancy, its deportment and its savoir-vivre.' [100]

[100] Deville, op.cit. p.21.

386

rehearsing a grand theatrical gesture feigning exhaustion (caused no doubt, as Giedion would assert, by her "confusing" environment). 36/f.8

36 Ibid., pp. 386–7.

All of these figures, which, like the "canvases" of "coyly posed nudes" of the nineteenth century relegated by modern museums to their "cellar," 37 as Giedion writes, escape their spatial confinement by leaping from the textual frames of the surrealist interiors. Here, the composite character of ornamentation addresses not only the hybrid composition of individual figures but the construction of spaces composed by surrealist montage that thrive in the creation of correspondences between divergent ontological conditions. The surrealist interior serves as proof that the machine has created not simply an assortment of individual artifacts, but has solidified its presence into an overwhelming spatial system. This is the legacy of the all-encompassing cycle of adornment, Giedion's *Ausschmückung*; that even after the eclipse of ornament, it survives as an interiorized, organized system.

37 Ibid., p. 353.

Giedion's archival research on nineteenth-century adornment proves that the "cellar," a veritable museum of past ornamental curiosities, becomes the research space par excellence of the psychoanalytically inclined architectural historian. Like the surrealist artist in search of figures from his "childhood memories," Giedion feels compelled to descend to this dusted archive and unearth a treasury of nineteenth-century statuettes and bric-a-brac only then to re-bury the same relics as funereal ornaments within the pages of his voluminous tome. His chapter on "Die Mechanisierung der Ausschmückung" is as much an account of the mechanization of adornment as it is a history of the adornment of mechanization; it describes not only the proliferation of mechanization through a series of functionalist implements but also the machine's impending aphasia caused by the

231. Sarah Bernhardt's Studio. 1890. (The Decorator and Furnisher, *New York*, 1891)

Toward 1880, as the confusion of feeling on the Continent is reaching its igh mark, the interiors of ruling taste dissolve into endless details and nuances, he sense and nonsense of which is a closed book to later generations.[101] This s also the time when the *confortable* becomes a drift of cushions.

Thus in the last decades, the authority of the upholsterer was ever increasing. Ie was the man to gather superficially loose ends. He provided oil paintings nd their gold frames for a middle class unable to afford originals. He arranged till lives from the bric-à-brac of a mechanized past. *Décorations mobiles*, the 'rench of 1880 called these strange compositions that were set up with an air f casualness on tables or chairs.[102] Cushions and heavy draperies completed he effect.

Here too the Surrealists tell what was taking place within. In one of his houlish *collages* of the 'Lion de Belfort,' [103] Max Ernst portrays the process by hich furniture was attacked. There, in semidarkness, a *confortable* stands upreme, with its *frange royale* sweeping to the floor; in it is posed one of those unmbiguous nudes which hung in the fashionable salons of the period. The beauty

[101] Cf. chapter 'Grammaire de l'Ameublement' in Henri Havard, *L'Art dans la maison*, nouv. ed., 'aris, 1884.

[102] How deeply such custom was rooted in the times may be gauged from its treatment by that eminent cholar Henri Havard, to whose authority we have so often referred. His book on the interior decoration f his day, *L'Art dans la maison*, contains a full-page reproduction of a *fauteuil* garnished with such a *écoration mobile.*

[103] Max Ernst, *Une Semaine de bonté ou les sept éléments capitaux;* Cahier: Le Lion de Belfort, Paris, 934.

387

hypertrophy of the ornament it compulsively kept churning out. Adornment then has a double function in modernity: it not only acts as an additional support, shielding the "command" of mechanization in modern culture and its transition from the aesthetic sensibility of the nineteenth century, but also serves as an historical vestige of that very power in the moment the machine's functional sovereignty becomes obsolete.

Media Machines and Not-So-Utopian Futures, c. 1957
Daniel A. Barber

Daniel Barber is Assistant Professor of Architecture at the University of Pennsylvania School of Design.

In the face of environmental pressures, architects and consultants are increasingly aggressive in their attempts to refine strategies of energy efficiency in building. This appears to be a worthy goal, and one of potentially historic consequence: without changes to our building practices, the effects of climate change could accelerate. Such endeavors—such a focus on the material efficiencies of the constructed environment—also resonate across historical phenomena, and allow for a new perspective on the architectures and ideas of the past.

What becomes clear, as a first principle, in examining these histories, is that the imperative for material efficiencies has often been accompanied by specific immaterial effects—cultural interventions that allow for a different relationship between natural and social patterns. Indeed, the cultural changes being solicited by the effects of climate change are much more dramatic, and significantly more of a challenge, than the technological prospects for reducing carbon dependence. This immaterial realm, where decisions about designing or building are examined not only for their reduction of fossil throughput but also for the new cultural patterns and forms of collectivity they imagine, awaits more specific elaboration, and more specific integration into the historical and theoretical discourse of architecture.

The image, broadly considered, is a primary site for this elaboration. Over the past century or so, images produced in an architectural context—diagrams, photographs, charts and graphs, plans, sections, and elevations, and, more recently, renderings—have been a central discursive site for analyzing and absorbing the impact of changes to the natural world. It is in part through this visual discourse that knowledge about the entanglement between human and biotic systems has crossed disciplinary and professional boundaries. Architects produce images, and images contain ideas that speak far beyond the constraints of the disciplines.

Machines produce images, too. In the 1950s, a number of machines were developed in order to more precisely understand the relationship between a building and its climatic conditions. Some of these machines were designed to produce specific material effects in buildings. Contingent historical events, such as the exploitation of oil in the Middle East, compromised the impact of these effects, as buildings were increasingly seen to be able to produce their own climate, isolated from the exterior and generated through the burning of fossil fuels. As a result, many of the devices that had been built to explore methods of making a building more energy-efficient were directed instead towards exploring the complexities of the experience of interior space, and focused on producing images that could imagine new conditions for the human. These are the machines that will be discussed below.

Mark Poster recently offered an important clarification on these terms, articulating a distinction between media machines and mechanical machines. "Media machines," he writes, "act on the components of culture, not nature (if that distinction may still be employed), affecting human beings in a way very different from mechanical machines." 1 Rather than processing materials, media machines process ideas. Poster at once affirms the material/immaterial distinction with which I began, and also indicates, if largely parenthetically and symptomatically, that the images that media machines produce operate precisely on the distinction between human and natural systems, and suggest the evaporating capacity to convincingly deploy that distinction. At stake here is the role that ideas about architecture have played in coming to terms with the entanglement of human and natural systems,

1 Mark Poster, "An Introduction to Vilém Flusser's *Into the Universe of Technical Images* and *Does Writing Have a Future?*," in Vilém Flusser, *Into the Universe of Technical Images*. Trans. Nancy Ann Roth (Minneapolis: University of Minnesota Press, 2011): pp. ix-xxvii; here p. x.

f.1 "A Not So Utopian Future ..." as illustrated in Eric Hodgins's "Power From the Sun."

and the images that have emerged to offer some clarity in this regard. 2 The design of machines that produce images also generates new conceptions across the nexus of the human and the natural world, and opens up access to histories and archives heretofore difficult to see.

To begin with the image of a machine: f.1 this rendering of "A Not So Utopian Future" accompanied, as a side bar, "Power from the Sun," an article written by the *Fortune* editor Eric Hodgins in 1953 on the potential of solar energy. 3 This image, and many others like it, and indeed many experiments in the architecture of solar energy, are an early indication that, although we tend to think of the American 1950s as a decade of endless economic expansion and consumer growth, of increased

2 Timothy Morton, *The Ecological Thought* (Cambridge, MA: Harvard University Press, 2012); see also Morton, "Architecture without Nature," *tarp: Architecture Manual*, 3 (2012), pp. 20–5.

3 Eric Hodgins, "Power from the Sun," *Fortune*, 48, no. 9 (1953), pp. 130–5, 184–94.

productivity and the unrelenting exploitation of resources—as the great acceleration, in short—voices of caution were raised even then about the relationship between social behaviors and what appeared to be the given condition of resource stocks, and of concern about how the collective behaviors of humans impacted the environment. [4]

These anxious voices were focused on gaining increased knowledge about those environmental conditions, and using that knowledge to better intervene in them. The best known of these voices today is that of M. King Hubbert, a research scientist at Shell Oil. In the late 1940s, through careful calculations involving industrial equipment, cost and profit analyses, and geological testing, Hubbert outlined the "[m]athematical relations involved in the complete cycle of production of any exhaustible resource". [5/f.2] Today we call it Hubbert's Peak—or Peak Oil—though he referred to this slim period when fossil fuels were readily available as a "pip" in the historical development of human affairs, "a transient and ephemeral epoch in the longer span of human history;" based on an assumption that fossil fuel sources would soon be replaced by "water power and solar radiation." [6] A related analysis, by a research scientist at Gulf Oil named Eugene Ayres, plays this out a little more carefully. In images he produced for distribution to the oil industry, Ayres attempted to suggest that different levels of research investment into renewables would lead to different possible futures. [7]

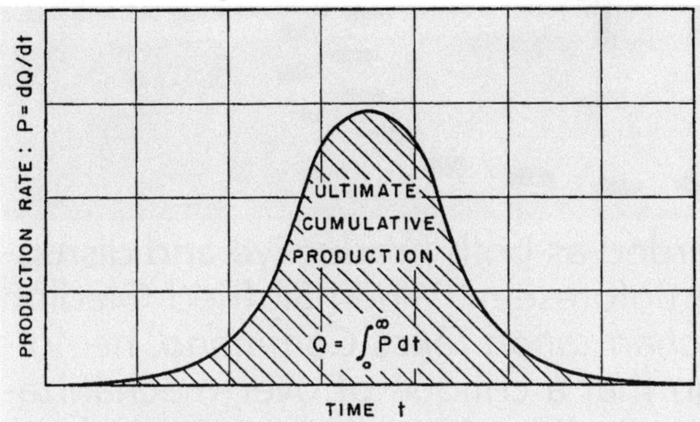

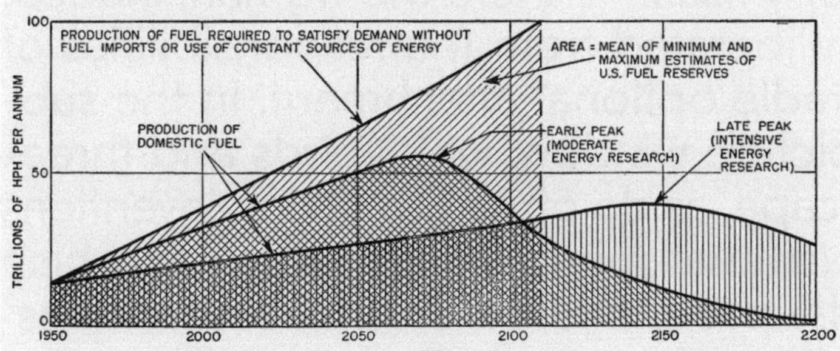

FIG. 5. Some possibilities in our future energy picture.

There were many other expressions of caution, gentle criticisms embedded in drawings, and images of possible futures that offered a check to the promise of progress, however it might have been envisioned. In another image of another imagined machine, the "Faucet Kitchen" drawn by the marketing team at the Schaible Company, a manufacturer of metal filters for kitchen sink drains, a different sort of future is envisioned. [f.3] Derisive to the point of absurdity of the many potential improvements technology was seen to bring—the "seeds and chemicals" and the "pre-dehydrated food processor"—the faucet kitchen indicates

4 Daniel A. Barber, *A House in the Sun: Modern Architecture and Solar Energy in the Cold War* (New York: Oxford University Press, 2016).

f.2 M. King Hubbert, "Mathematical relations involved in the complete cycle of production of any exhaustible resource," from *Nuclear Energy and the Fossil Fuels* (1956), based on a report from 1949; and Eugene Ayres, "Some Possibilities in Our Future Energy Picture," from *Energy Sources: The Wealth of the World*, 1952, based on a 1948 report.

5 M. King Hubbert, *Nuclear Energy and the Fossil Fuels* (Houston: Shell Development Co., 1956), fig. 11, based on a report from 1949, see Hubbert, "Energy from Fossil Fuels," *Science*, 109, no. 2823 (1949), pp. 103–9.

6 M. King Hubbert, "Exponential growth as a transient phenomenon in human history," in Margaret A. Strom (ed.), *Societal Issues, Scientific Viewpoints* (New York: American Institute of Physics, 1987), pp. 75–84; here p. 76.

7 Eugene Ayres, "Major Sources of Energy," in *Addresses and Reports Delivered at the Twenty-Eighth Annual Meeting, Chicago, Illinois, November 8 to 11, 1948* (New York: American Petroleum Institute, 1948), pp. 109–44.

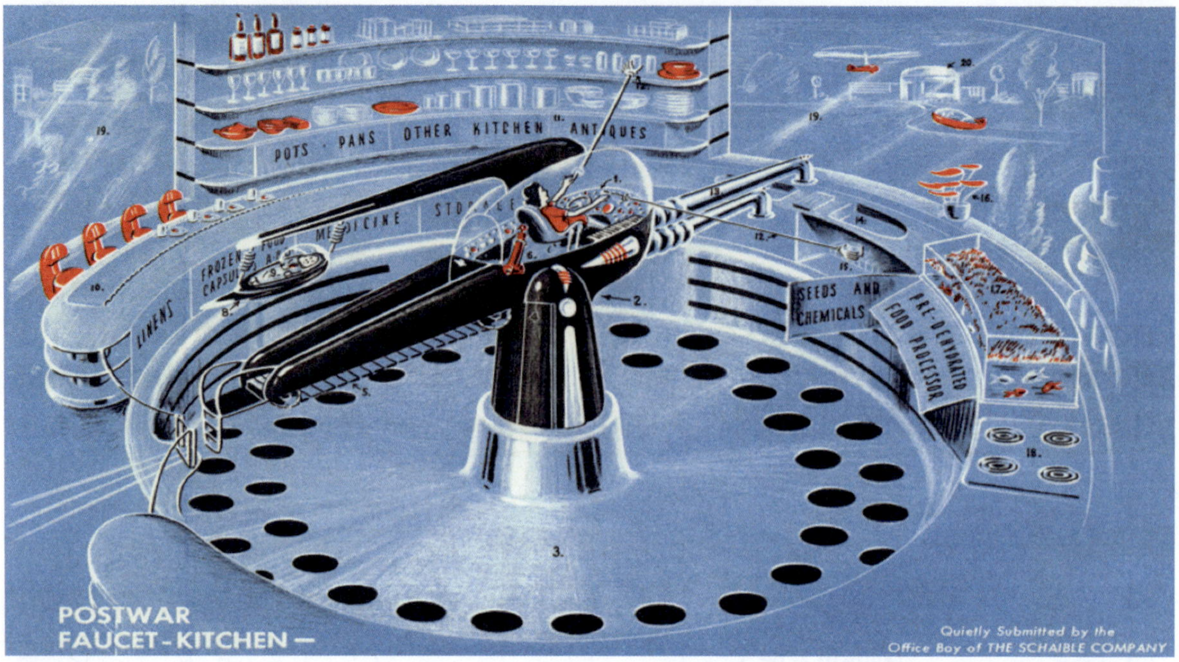

f.3 Schaible Company (manufacturer of sink drains), *Kitchen of Tomorrow*, advertising brochure, 1945.

that technologies were regarded as both productive and disruptive, and as portents of the unforeseen. When Sigfried Giedion published this image in *Mechanization Takes Command*, he noted that "[i]t is a healthy sign that a critique of over-mechanization starts from within industry itself." 8 Here the woman, indeed, is fully in charge, and even liberated from the responsibilities of childcare by the rocking cradle optional attachment. In the suburban landscape out the picture window, whirlybirds and three-wheeled cars dot the landscape, as do other Dymaxion inventions such as the cylindrical house that mirrors the one out of which the viewer is looking. "Pots, pans, and other kitchen antiques" rest in the background—still of some value, apparently, as the mechanical hand is taking a bottle off the shelf.

8 Sigfried Giedion, *Mechanization Takes Command: A Contribution to Anonymous History* (New York: Oxford University Press, 1948), p. 580.

The not-so-utopian house imaged in *Fortune* plays this out on slightly different terms. "[T]he day is coming," the author wrote, "when population pressures will make wheat fields and cattle ranges luxuries of a dear dead past." This house was the solution, a survival system for a family of four. It had south-facing windows to absorb the heat of the sun in warming the interior; and also used "solar radiation to grow its own food, in the form of algae on its roof." In the "workshop"—a sort of elaboration on the mechanized postwar kitchen—garbage was burned to charge the algal suspension and keep it growing; this algae was filtered, heated, and then could be poured out as "a concentrated sludge ... a dark-green paste with a pleasant grassy odor ... [and] enough fresh organic matter ... to supply the entire protein requirements for the family." 9 "We will come to rely," the article concluded, on "the internal metabolic adjustments by which we shall subsist contentedly ... on hydrolysed sawdust and predigested, vitaminized algae." 10 The key word here is

9 Eric Hodgins, "Power from the Sun" (see note 3), p. 134.

10 Ibid., p. 194.

"contentedly"—metabolic adjustments and changing conceptions of the psychological and biological makeup of humans will adapt, it was proposed, to a parched earth. Here, the image, imbued with a mixture of irony, paranoia, and caution, suggests the production of a new kind of subject; or, at least, schematizes a relationship between technology, humans, and the environment in which the production of new subjectivity and new physiology accompanies the elaboration of built efficiencies.

Though the house-as-machine imaged in *Fortune* was speculative, if not in fact imaginary, it emerged in relationship to a number of machines, images, and ideas that sought to increase professional knowledge of the potential state of these unknown futures. Innovations in building design and technology, it was supposed, would allow humans to live differently in future conditions, and media machines were essential to producing the future conditions by which human and natural systems would interrelate. Among the most engaging of the machines focused in this period on architecture and the environment was the thermoheliodon. f.4 Designed and built by the Hungarian émigré architects Victor and Aladar Olgyay in 1956 at the Princeton Architectural Laboratory, the thermoheliodon offered a relatively precise picture of the environment in which a building would be placed, so as to render more precise the methods that could inform that building's design.

The device followed on a large number of heliodons designed and built at universities and research centers around the world since the mid-1930s. The heliodon is a relatively simple device: a sunlamp is calibrated along the vertical calendar to provide the seasonal height, and then the building is placed on a platform angled according to latitude. The platform can spin to simulate diurnal patterns relative to the sun's location. [11] In the early 1950s, a group of students at Columbia University calling themselves "The Form and Climate Research Group" used a heliodon there, as well as a simple wind tunnel that they built, to test buildings according to a relatively unsophisticated set of climatic parameters. [12] Other heliodons were built at the University of Kansas, Princeton, and at a number of the tropical building research stations in former British colonies. [13]

The Olgyays, though not well known today, were recognized in the postwar years for their two books, *Solar Control and Shading Devices* (1957) and *Design with Climate: A Bioclimatic Approach to Architectural Regionalism* (1963), both of which entered into the curricula of architecture schools and sat on the reference shelves of design firms. They built the thermoheliodon to improve on the heliodon by offering a much more precise

11 One of the first in the U.S. was built by the planner Henry Wright at Columbia University's Graduate School of Architecture. See A. F. Dufton and H. E. Beckett, "Orientation of Buildings: Sun Planning by Means of Models," *Journal of the Royal Institute of British Architects*, 38 (1931), p. 50; and G. Manley, "Microclimatology—Local Variations of Climate Likely to affect the Design and Siting of Buildings," *Journal of the Royal Institute of British Architects*, 56 (1949), pp. 317–23.

12 "The 'Form and Climate' Research Group," *Interiors*, 112, no. 12 (1953), pp. 52–3.

13 George Atkinson, "Building in the Tropics: Research into Housing in Tropical Countries, especially in the Commonwealth," *Journal of the Royal Institute of British Architects*, 57 (1950), pp. 313–9.

f.4 Victor and Aladar Olgyays' thermoheliodon in the Princeton Architectural Laboratory, with explanatory drawing, c. 1957. → 160/161

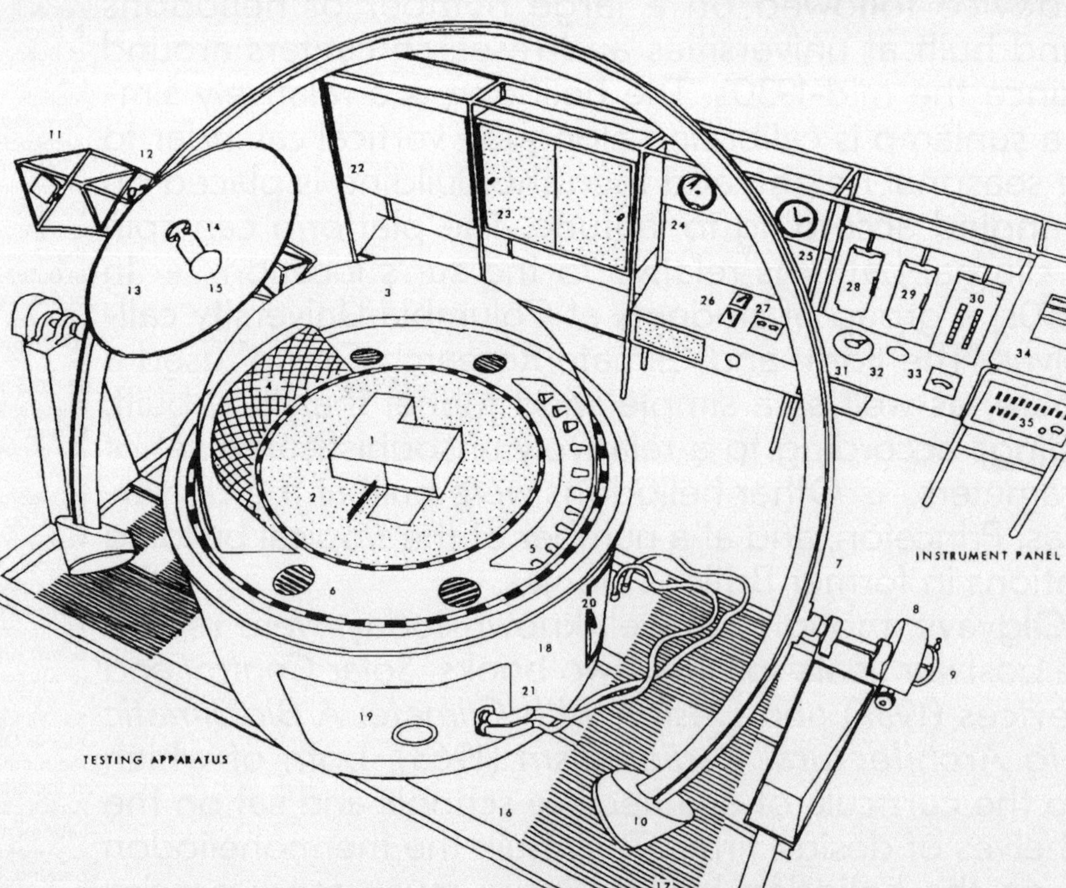

EXPLANATORY DRAWING OF THE THERMOHELIODON

DESCRIPTION OF THE THERMOHELIODON

The laboratory machine consists of two major elements: the testing apparatus and the instrument panel. The description of its parts (see illustration on page 7) is as follows:

1. Experimental model constructed with architectural scaling 1' = 3/4", and thermal scaling of 1/6th of prototype.

   For exterior temperature measurements thermocouples, for interior heat observations thermistors are attached to it.
2. Experimental area of 4' in diameter consists of a sunken pan filled with soil native to locale being tested to insure proper thermal relations with ground. The pan rotates to give any desired wind orientation.
3. Plexiglas dome to cover experimental area and allow recirculation of the air for heating and cooling purposes.
4. Air inlet constructed from strips of corrugated plexiglas for least interference with radiation effects. The inlet is designed to direct air over the experimental area. Steel mesh screening in the throat of the inlet insures balanced airspeeds.
5. Air outlet, receiving air for recirculation. Ten heating coils operated in groups of two simulate temperature fluctuations during the test "day."

6. Grills over four 300 watt incandescent lamps which by reflection from the dome's inner surface simulate diffuse sky radiation. The surface of the outer ring of the experimental surface is covered by neutral gray linoleum.
7. Latitude ring forming track for sun's path, variable according to latitude.
8. Protractor for positioning latitude.
9. Crank for tilting latitude ring.
10. Counterweight.
11. Month bridge allowing compensation in sun's path for seasonal variations. Adjustments are calibrated in 10-day intervals.
12. 1/75 HP electric motor geared to drive sun around the latitude ring. Path of 180° run (12 hours scaled to 1/36) takes 20 minutes.
13. Polished aluminum parabolic reflector of 4' in diameter to produce parallel rays from radiation source.
14. Incandescent 5000 watt bulb for simulating sun's radiation.
15. Small hemispherical aluminum reflector to intercept non parallel direct radiation and redirect energies to parabolic reflector. Not shown in the drawing is the equalizing grill before the large reflector.

16. Understructure of the testing apparatus built of steel I sections.
17. Iron cat-walk.
18. Cylindrical base of experimental area. The whole structure rotates on cast iron rollers to allow for any desired sun orientation.
19. Grill to allow access to one HP motor with variable pulleys to achieve wind velocities from 50 to 1000 ft./minute. Most tests are run at 440 rpm (5mph). A 24" fan circulates the air through a wire lath and asbestos plaster throat toward the inlet.
20. Exhaust door for expelling warm air and bleeding in room temperature air for cooling purposes.
21. From the cold junction and terminal strips (not shown in drawing as it is on the other side of the base) the temperature measuring wires enter into a flexible plastic tube which allows rotation of the base. The wiring from here is connected to the instrument panel through a conduit under the floor.
22. Chalk board.
23. Compartment for instruments.
24. Ambient temperature and relative humidity recorder.
25. Clock geared to run at scaled time (one hour equals 100 seconds).

26. Two 8-point switches for thermocouples.
27. Potentiometer to measure the emf developed by the thermocouples.
28. Main switch for the artificial sun's 50 amp current supply.
29. Main switch for low current supply.
30. Control switches for heaters, indirect radiation, fan, sun-motor drive, and for instrumentation.
31. Powerstat for varying voltage and consequent intensities of sun radiation.
32. Variac for varying intensities of diffuse radiation.
33. Galvanometer to register air temperatures under the dome.
34. Automatic recording instrument adjusted to read at 2-second intervals thermistor relayed fluctuations of temperature produced inside the model house.
35. Balancing panel for thermistors.

Page 13 illustrates the plan, views, and section of the Thermoheliodon in geometrical projection.

accounting of the site's climate. The Olgyays had been brought to Princeton in 1953 to work on an analysis of shading systems for the aluminum manufacturer ALCOA, as part of a broader research project on metal curtain walls. The Princeton Architectural Laboratory, built in 1948 by Jean Labatut, was the site of these experiments—the Olgyay twins, hampered by their poor spoken English and by perpetually temporary, grant-based contracts (they were officially "Research Associates in Architecture," with very limited teaching duties) focused their activities at the Lab.

The components of the thermoheliodon precisely express how climate was defined in the period: the arc of the sun was carefully calibrated; there was also an adjustable screen for refining wind direction, and a shallow pit in the middle, where soil from the building site was to be placed; humidity was controlled by moisture input from jets on the right; and all of this was sealed in a dome that could approximate some larger-scale atmospheric patterns. Compared to the heliodons just described, significantly more detailed models of the climatic world became available here to the designer.

Referred to as a "Laboratory Machine for Testing Thermal Behavior of Buildings," it was developed so the Olgyays could better conceptualize and calibrate what they came to call the "comfort zone"—the thermal state, as they defined it, "wherein the average person will not experience the feeling of discomfort." [14] The zone was modeled on a number of bioclimatic charts—similar to many of the physiometric charts of the early twentieth century, but distinct in the clarity of the specific condition ideal for human habitation. f.5 In this image, the y-axis is temperature; the x-axis is humidity. The upper dotted line, angled as it responds to both temperature and humidity, indicates a limit beyond which there is danger of sunstroke; the lower dotted line simply indicates freezing. The centerline is speculative, suggesting how a shading system cuts through these extremes and neutralizes them—and it sits, as one can see, at the bottom of the comfort zone.

These are just some of the images that the thermoheliodon helped to produce. Based on climate analyses, data from a given area could be placed on this graph, and analyzed. These data points would be translated into drawn shapes for each month, and arranged across the graph to show their relative co-extension with the comfort zone in a non-manipulated state. Other drawings interpreted this chart as a "timetable of climatic needs," with the dark areas in the center showing the over-heated periods that require a focused shading approach. Based on such an analysis, a basic building shape could be developed, with

14 Victor Olgyay, *Design with Climate: Bioclimatic Approach to Architectural Regionalism* (Princeton, NJ: Princeton University, 1963), p. 18.

material and shading differences across each façade so as to best engage with the specifics of the building's climatic surround. A building model was then constructed and placed in the center of the thermoheliodon device and subjected to a number of tests, based on what was a relatively remarkable amount of data about a given building site.

At the same time it is worth noting that the thermoheliodon, on its stated functional terms, did not really work. The problem had to do with the thermal capacity of materials: the model buildings that were inserted into the device were tested for shape and orientation, but the interior climatic conditions—the comfort zone itself—could not be adequately monitored because of the difficulty of scaling up the thermal properties of materials. A small brick operates very differently, on thermal terms, than does a large brick. Indeed, a significant portion of the report they submitted to the National Science Foundation (NSF) focused on "Scaling Criteria for Heat Transfer in Model Experiments." [15] Their attempts to take into account materials, humidity, and heat exchange read almost as an extended lament expressed in calculus—a longing for a more direct means to predict a building's performance. When seeking additional funding, they proposed to build identical test houses in Princeton, Montreal, and Los Angeles, and to maintain constant data analysis of these three sites, triangulating and adjusting their calculative matrix according to the recent historical record. Their funding was not renewed.

15 Victor Olgyay, *Report on the Thermoheliodon: Laboratory Machine for the Testing of Thermal Behavior through Model Structures* (Princeton, NJ: Princeton School of Architecture, 1957), pp. 39ff.

The thermoheliodon was at the center of a diverse body of images, machines, and buildings that attempted to recast the role of architecture in examining and understanding the relationship between humans and nature in the period. The Olgyays designed a number of houses in the Princeton area, and also consulted with well-known mid-century architects to provide shading systems for their designs. Among many other projects, they worked with Josep Lluis Sert on Peabody Terrace in Cambridge, MA; with Bernard Zehrfuss, Pier Luigi Nervi, and Marcel Breuer on the UNESCO building in Paris; and with Walter Gropius and The Architects Collaborative (TAC) on plans for the University of Baghdad. Aladar also collaborated with the engineer Maria Telkes on developing a prototype of the "Solar Wall" for the Curtis Wright Corporation. [16]

16 Aladar Olgyay and Maria Telkes, "The Principle of the Solar Wall," *American Artisan* (August 1949). Information on the Olgyays' collaborations comes from the Victor Olgyay Collection, Special Collections and Archives, Design Library, Arizona State University.

With the thermoheliodon, as with the not-so-utopian house, the physical capacities of the machine are not its most performative components—the immaterial, more than the material effects, are of interest. The experiments undertaken with the device and the images that it produced allowed for new conceptions of the

f.5 The "Schematic bioclimatic index" from Victor Olgyay, *Design with Climate*, 1963.

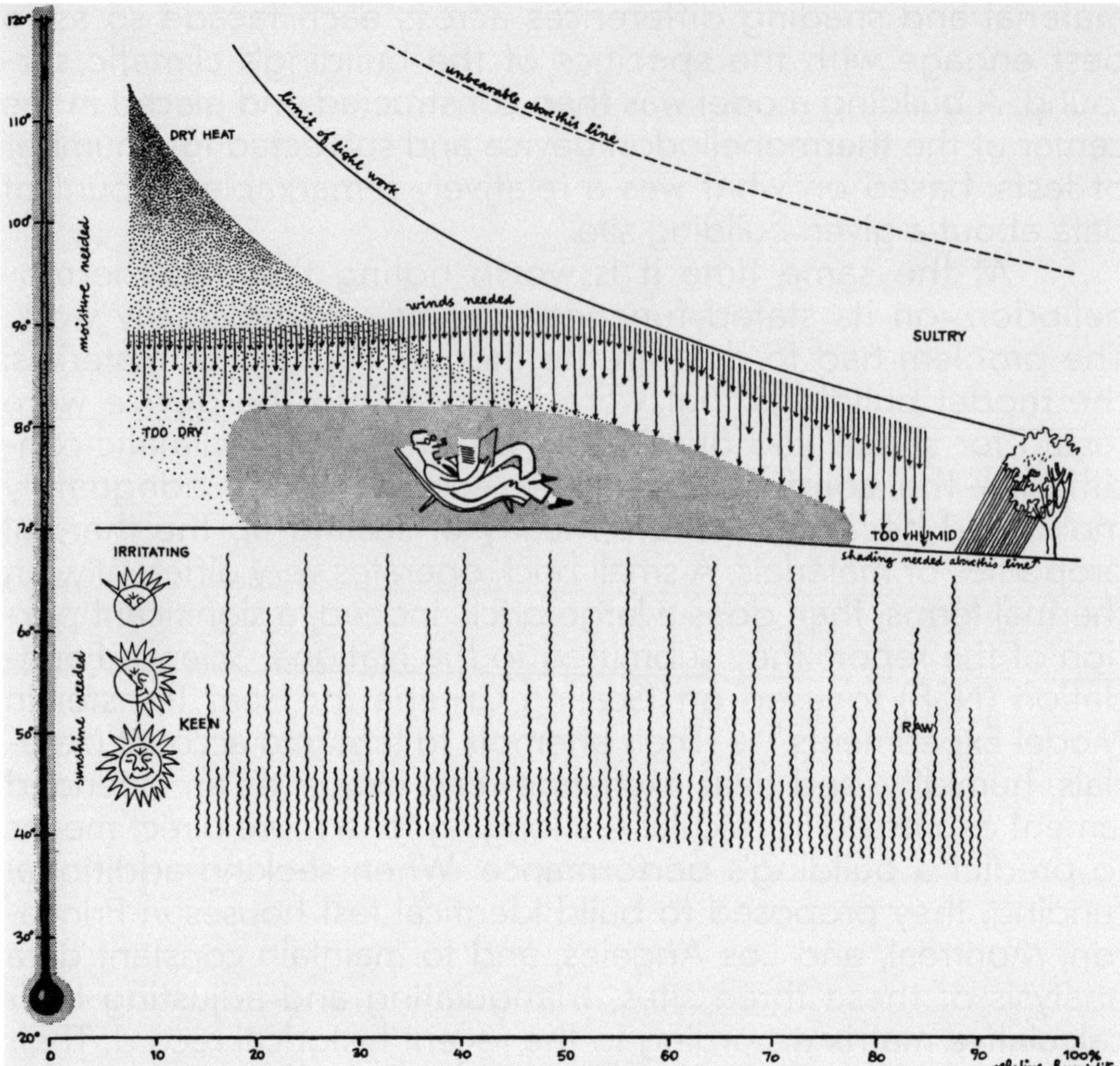

46. Schematic bioclimatic index.

human. Indeed, the thermoheliodon was itself something of a test case for more precisely understanding how increased technical knowledge—about the climate system, about architectural methods to engage it—could inform new desires, and lead to new ways of living. Some of the images that it produced constitute a new type of diagram, one that operates as an expressive image of technological knowledge—an image that places "humans" explicitly at the center of a complex arrangement of "natural" and "architectural" elements, as a means to operate upon all three of these concepts simultaneously.

The Olgyays' "schematic bioclimatic index" clarifies the extent to which the thermoheliodon and its attendant diagrammatic practices was a biopolitical project, an attempt to enact a new mode of being, for better or worse—a mode we can see today as shot through with the anxiety of the not-so-utopian. f.5 The image of the human in figure 5 is protected, sheltered in an ideal, abstracted space—it could well be Aladar Olgyay, as he was always photographed with a pipe—relaxing on a chaise longue, reading the newspaper and completely at ease; without irritation, and apparently beyond the reach of any of the potentially threatening climatic conditions surrounding him. Temporally and spatially, he is a stable, protected figure—solid in the experience of well-designed space; consistent across changes in the elements that the past, present, or future can bring. In 1957, so this quasi-technical image seems to suggest, the project of climatic simulation was to produce a condition of stasis—a stasis that was, importantly, rendered as such through the intervention of carefully considered, machine-tested architectural methods.

Though the thermoheliodon was designed and built long before the specter of climate change cast a shadow on the prospects for unlimited economic expansion, these images of and aspirations for a static human condition are implicated in and inserted into a set of relationships that have, since the mid-twentieth century, attempted to assess the possible consequences of human behavior as it relates to the state of the global ecosystem. The Olgyays' bioclimatic analyses and their engagement at conferences with physiologists, biometeorologists, and ecologists clarifies, again, how the thermohelidon's image effects helped

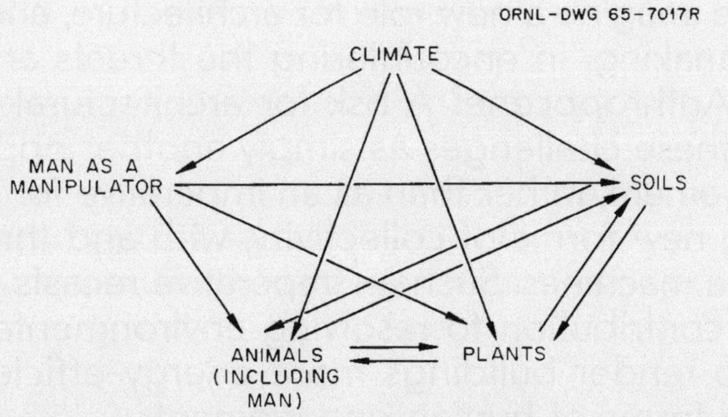

Fig. 1. An ecosystem is an integrated complex of living and nonliving components. Each component nfluenced by the others, with the possible exception of macroclimate. And now man is on the verge of rting meaningful influence over macroclimate.

f.6 George van Dyne, "An ecosystem is ..." from *Ecosystems, Systems Ecology, and Systems Ecologists*, 1966.

to restructure the distinction between humans and nature just as the epistemological significance of that distinction was beginning to dissolve. The concept of "the environment" as a site for human anxiety was taking root, and the notion of the ecosystem, being developed in the biological sciences, began to express this re-conception most clearly. **f.6** There was a growing recognition of the complexity of human agency—in the diagram in figure 6, drawn by George van Dyne as one of the early explanations of the ecosystem concept, "man" is positioned both as part of the animal world and "as manipulator;" i.e. as a force of nature itself, biological and eventually geological, that here begins to be conceived of as not subject to environmental conditions, but as having a primary role in transforming them. As Van Dyne noted in 1966: "now man is on the verge of exerting meaningful influence over macroclimate" while persisting, all the same, in "his" animal or biological state. [17]

This complex consideration of human and environmental agency, developed in different contexts at least since the mid-nineteenth century, has reached something of an apex in the contemporary framework of the Anthropocene — the seemingly new epoch, "the human era" in which we all have been living for some time, defined by a perspective that sees human activities as structuring geophysical and geological systems. [18] The positioning of humans and environments in systems models also begins to suggest a re-positioning of the species: rather than protected, stable, and static, the human is vulnerable, and subject to risk — and also to hubris, mishap, and misconception. Entangled, rather than in control. Media machines in the 1950s produced images that help to frame species-existential threats of climate change on a broader historical horizon. At stake, in the end, is the normative approach to the relationship between understanding and agency that persists. [19] What other conceptions can emerge, as we imagine a new role for architecture, and for speculative image-making, in encountering the threats and opportunities of the Anthropocene? A risk for architectural discourse is to approach these challenges as simply another opportunity for design engagement, rather than as an imperative for speculation, for imagining new forms of collectivity, with and through architectural-media machines. Such an imperative recasts the promise of the field's contribution to resolving environmental problems: not merely to render buildings more energy-efficient but also to restructure forms of human engagement.

17 George van Dyne, *Ecosystems, Systems Ecology, and Systems Ecologists* (Oak Ridge, TN: U.S. Atomic Energy Commission, 1966), p. 3.

18 See, for example, George Perkins Marsh, *Man and Nature; or, Physical Geography as Modified by Human Action* (Seattle: University of Washington Press, 2003 [1865]). On the Anthropocene, see Eileen Crist, "On the Poverty of Our Nomenclature," *Environmental Humanities*, 3 (2013), pp. 129–47.

19 Wendy Hui Kyong Chun, "On Hypo-Real Models or Global Climate Change: A Challenge for the Humanities," *Critical Inquiry*, 41, no. 3 (2015), pp. 675–703; here p. 677. See also Ulrich Beck, *Risk Society: Toward a New Modernity* (Thousand Oaks, CA: Sage, 1992).

The Model as a Machine: Frei Otto's Architectonic Experimental Systems

Georg Vrachliotis

"People say that machines, the tools of development, kill fantasy. That's not the fault of machines, however, but of man himself, who builds machines that are neither adaptable nor rational: machines that are lacking in fantasy." 1 Thus wrote the young German architect, Frei Otto, in his concise, manifesto-style essay "Imagination et architecture: essai d'une vision d'avenir," which was published in the French journal *L'Architecture d'Aujourd'hui* in 1962. Otto illustrated his essay not with photos of realized projects but with watercolor drawings and sketches in Indian ink, in which intriguing membrane roofs and tent-like structures could be seen to span residential developments, landscapes, or even entire self-contained cities. In juxtaposing factors ostensibly worlds apart—the unfettered imagination; constructive or machinic potential—Otto put his finger on what would prove to be a core tenet of his architectural practice: the abiding aspiration to strike an effective and fertile balance between creative vision on the one hand, and the rationale of technological constraints on the other. **f.1 a–c/f.2**

Georg Vrachliotis is Professor for the Theory of Architecture at Karlsruhe Institute of Technology (KIT).

1 Frei Otto, "Phantasie und Architektur," *db deutsche bauzeitung*, 69 (1964), pp. 543–4; here p. 543; (French original: "Imagination et architecture: essai d'une vision d'avenir," *L'Architecture d'Aujourd'hui*, 102 (1962), pp. 89–93. This and all further translations by Jill Denton.

f.1 a–c Soap-film model of the dance pavilion, Federal Horticultural Show, Cologne, 1957.

In the early 1960s Otto was still at the start of his career. Yet already he had laid the theoretical and practical groundwork for a modern minimalist reading of the tent. 2 His early dynamic tensile structures, seemingly suspended in mid-air, epitomized the longing of his generation for an open, enlightened society, and lent a new face to the young German Federal Republic. The ideal of eternal, monumental, and prestigious architecture was swept aside by a striving both to perfect construction by minimalist means and to inscribe mutability and ephemerality in the interface of architecture and engineering—artistically, technically, and socially. Otto's reflections on the relationship between machines and fantasy, cited above, were first and foremost a critique of the architectural concepts prevailing still in the immediate postwar period. The "fantastical line" had run dry, he noted, owing to the intently rational formal idiom of classical modernism. 3 Invoking fantasy, by contrast, would call to mind the experimental architecture and optimistic narratives of Expressionist movements of the 1920s, the poetic force and literary ingenuity of Paul Scheerbart and Bruno Taut, in short, the Expressionists' celebration of the spiritual dimension of worldly experience.

2 To name but a few examples, Otto's modern tent-like constructions included a four-point tent for the Federal Horticultural Show in Kassel (1955), a star-shaped four-point tent for the dance pavilion and entrance arch at the Federal Horticultural Show in Cologne (1957), the so-called humped pavilion for the International Architecture Exhibition (IBA) in Berlin (1957), and the pointed tent for the Swiss cantonal exhibition in Lausanne (1964). Cf. Conrad Roland, *Frei Otto—Spannweiten: Ideen und Versuche zum Leichtbau* (Berlin: Ullstein, 1965).

3 Otto, "Phantasie und Architektur" (see note 1), p. 543.

It was therefore no accident that Heinrich Klotz—shortly after founding the German Museum of Architecture in Frankfurt in the mid-1980s—gave center-stage in one of the first exhibitions there to Otto's lightweight structures. This was prompted by nothing less than Klotz's own injunction to once again expose "modernism's other root" given that the "reductive geometry of the Bauhaus," which had long reigned supreme as "modernism's normative benchmark," had patently had its day. [4] Otto had made his mark, Klotz felt, by confidently exploring and exploiting "a promising aspect of modernism that *Neues Bauen*, Le Corbusier, and Gropius had overlooked: the art of engineering." [5] Yet, although Klotz paid a great deal of attention to Otto's architectural projects, he neglected one crucial point: the originality of Otto's design lies in the originality of his design process. Impossible, therefore, to adequately demonstrate the presumed development of modernism's other root by recourse to built projects alone. Far more pertinent, in this regard, is an investigation of the intricate workings of tools and apparatuses and their specific operative coupling as an ensemble of things, which is to say, of the technical procedures and medial and social practices that underpin any order of knowledge and without which no building as outcome can ever be envisaged. [6]

4 Heinrich Klotz, "Vision der Moderne," in Klotz (ed.), *Vision der Moderne: Das Prinzip Konstruktion* (Munich: Prestel, 1986), pp. 9–26; here p. 10.

5 Ibid.

6 Cf. Gilbert Simondon, *Du mode d'existence des objets techniques* (Paris: Aubier, 1958), here in particular the chapter "L'invention technique: fond et forme chez le vivant et dans la pensée inventive," pp. 56–60.

Otto wrote his aforementioned essay two years after leaving the Ulm School of Design, where he had held the position of visiting lecturer for industrial construction from 1959 to 1960. Founded in 1953, the Ulm School of Design represented not only the perpetuation but also the institutionalization of the interwar Bauhaus tradition. However, the unity of art and technology propagated at the Bauhaus, above all by Walter Gropius, ceded there to a more analytical and methodological scientific concept of art and architecture, the visual expression of which took an abstract, namely diagrammatic, form. Implicit in this shift was a radical rethink of what exactly a machine might be. In publishing *Cybernetics; or, Control and Communication in the Animal and the Machine* in 1948, the mathematician Norbert Wiener had laid the foundations of a novel machinic realm in which the characteristics of a machine were taken to be not so much the mechanics of material itself (objects), as the mathematical control and, above all, the auto-control of abstract input and target variables. Wiener's book was quickly acclaimed in Ulm as the go-to work of reference, and it also proved pivotal to the School's theoretical articulation of its institutional role, in particular thanks to the philosopher Max Bense and the design theoretician Tomás Maldonado. 7 In consequence, the importance of creativity and the social imagination became subordinate to the artificial intelligence of electronic brains and thinking machines. Design processes were reduced to scientific procedures for problem resolution, which were calculated using a methodology comprised of circuit diagrams and feedback loops. Aesthetics too was stripped of sensibilities and personal expression, and read instead as a communications technology issue. The rigor of this approach rested in many respects on cybernetics' claim to be a universal science and hence on the supposition that the borders between object and subject, nature and culture could be overcome in order to arrive at a novel way of "thinking by modeling" and a superordinate method of scientific global analysis. 8

Young Otto did not join Ulm in singing the praises of techno-intellectualism, however understandable this would have been in light of general euphoria about progress at the time, particularly in postwar Germany. Instead, he pursued a line of inquiry that was not so much concerned with the imponderable nature of a universal science as with tackling the axiomatic terms of architectural practice. Likewise Ulm's highly theoretical claims and strictly scientific notion of design must have struck Otto as strange. 9 His endeavor to clarify the relationship between architecture, fantasy, and the machine must therefore be taken not only as a homage to the experimental architecture of Expressionism but also, so the hypothesis of the present essay, as a subtle critique both of the machine theory propagated at the Ulm School of Design and the cybernation of architectonic discourse therein implicit.

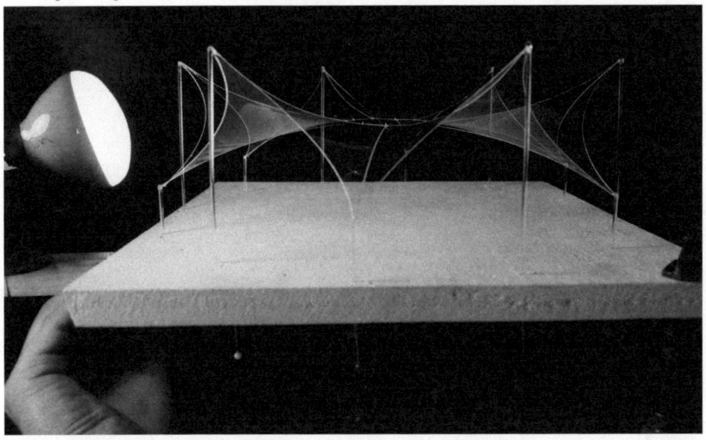

In 1964, Frei Otto was appointed to an institution founded in his behalf—the Institute for Lightweight Structures at the University of Stuttgart—where, in collaboration with architects, engineers, biologists, physicists, and artists, he quickly forged a world-class hub of interdisciplinary research. 10 It was in this period too that he won the competition for the design of the German Pavilion at EXPO 67 in Montreal. Together with the architect Rolf Gutbrod, Otto conceived an open exhibition landscape composed of spacious visitor terraces beneath a seemingly

7 Cf. Max Bense, "Kybernetik oder Die Metatechnik einer Maschine" [1952], in Bense: *Ausgewählte Schriften*, vol. 2: *Philosophie der Mathematik, Naturwissenschaft und Technik* (Stuttgart: Metzler, 1998), pp. 429–46.

8 Cf. Michael Hagner, "Vom Aufstieg und Fall der Kybernetik als Universalwissenschaft," in Hagner and Erich Hörl (eds.), *Die Transformation des Humanen: Beiträge zur Kulturgeschichte der Kybernetik* (Frankfurt/Main: Suhrkamp, 2008), pp. 38–71. See also: Georg Vrachliotis, *Geregelte Verhältnisse: Architektur und technisches Denken in der Epoche der Kybernetik* (Vienna: Springer, 2012).

9 "The technical and philosophical realms and ever greater challenges are so extensive now that the work of the architect is becoming increasingly cerebral. He thus easily becomes distanced from his work and loses the capacity to tackle it himself, single-handedly, as the sculptor does who carves stone, forms clay, cuts wood, or bends steel, and who by the labor of his own hands is able to give himself completely to his craft. ... If one senses that something is missing, one hides the fact in academic exchanges. The urge to explain is a fashion born of the misapprehension of science." Frei Otto, "Anmerkungen," *Baukunst und Werkform*, 8 (1955), pp. 721–2; here p. 721.

10 Cf. Daniela Fabricius, "The Spinner Experiment: Frei Otto and the Institute for Lightweight Structures," European Architectural History Network Meeting, Brussels, May 31 to June 3, 2012.

f.2 Nighttime view of the dance pavilion, Federal Horticultural Show, Cologne, 1957. → 170/171

f.3 a–b Wooden model in the wind tunnel, German Pavilion for the EXPO 1967 in Montreal, 1967.

free-floating yet meticulously structured tent-like roof. Once again, the Federal Republic of Germany had chosen to showcase its engineering skills in the form of an experimental lightweight structure, the acclaimed debonair flair of which made it a media darling the minute it opened. Soon, the term "Swinging Germany" was on everyone's lips. 11 Yet the apparent instantaneity of this fascinating structure belied its long and complex history: Gutbrod and Otto had designed not only the pavilion, but also all the requisite models, tools, measuring instruments, and visualization equipment for its design, planning, and implementation. Their venture was therefore nothing less than an apparatus-led rewrite of the cultural technics of design, the will to build revealed in an inventor's guise. And that, in a sense, is what made it a star attraction. f.3 a–b

11 Rudolf Leonhardt, "Swinging Germany," *Die Zeit*, May 12, 1967.

To calculate the enormous roof span Otto had turned to what is known as model statics, an empirical method commonly used in engineering to examine the interplay of external boundary conditions and internal force distribution by direct reference to scale models. To simulate the German pavilion's exposure to wind-power, for example, he used a simple wooden scale model and a wind tunnel. Small holes drilled in the model's surface and equipped with thin plastic tubing enabled variations in its resistance to air velocity and pressure to be precisely ascertained. Meticulous documentation of the results clearly demonstrated how wind-power would affect the pavilion's complex tensile roof. Models thus served Otto as a means not only to elaborate form but also to measure and chart the balance of forces engineered by variable configurations of material, load-bearing system, and construction method. As he wrote of one of his later inventions:

"the measuring table is a machine in which a test point can be moved in the coordinates x, y, z. A threaded spindle powered by an electric motor traces the trajectory of each coordinate. Rotations of the spindles are measured. The measured data are shown on an LCD monitor and can be printed in plain text or as punched tape. The punched tape can be fed into computers and automatic drafting machines. On the measuring table's drawing board, while coordinates are being measured, the ground plan of the model can be punched in card with the aid of an automatic needle." 12

12 Frei Otto, "Protokoll: Über die Arbeiten des Instituts für Leichte Flächentragwerke an den Modellversuchen und Auswertungen am Projekt Deutscher Pavillon, 1967, Weltausstellung, Montreal," mimeographed manuscript issued by the Institute for Lightweight Structures, March, 1967, n.p.

Frei Otto used such pioneering experiments in CAD (computer-assisted design) in particular for the construction of the

timber gridshell of the Multihalle in Mannheim. One might join Gilbert Simondon in simultaneously describing this apparatus-based extrapolation process as "concretization" and "differentiation." 13 The technical object is embedded in a technical ensemble that is essential to its deployment and without which it would be incomplete. In building physics, then, models and machines become equal players in the same experimental system. f.4/f.5

13 Simondon, *Du mode d'existence* (see note 6), pp. 34–5.

How radically the interaction of model, machine, and materiality was explored here is clear from the broad range of materials brought into play. By the late 1950s, Otto had established that mixing distilled water with a few drops of dish soap suffices to create extremely thin yet relatively stable soap films and that

f.4 Net model and overall model on the measuring table, German Pavilion for the EXPO 1967 in Montreal, 1965.

(as with a child's bubble wand) such soap films take the shape of any closed, bent-wire form briefly dipped into then removed from them. 14 If the bent-wire form describes a so-called space curve then the soap film likewise constitutes a three-dimensional arcuate surface. In the course of countless experiments, Otto observed that a membrane of this sort is primarily defined by its contours, which is to say, its highest and lowest points; and also that it has specific geometric and physical properties. The surface tension in a soap film is identical at all

14 Cf. Daniela Fabricius, "Capturing the Incalculable: Frei Otto's Experimental Models," in Sonja Hildebrand and Elisabeth Bergmann (eds.), *Form-Finding, Form-Shaping, Designing Architecture: Experimental, Aesthetical, and Ethical Approaches to Form in Recent and Postwar Architecture* (Mendrisio: Mendrisio Academy Press, 2015), pp. 49–63.

its points and in all directions, and this assures particular stability and efficiency, since the soap film takes the most economical form possible: the so-called minimal surface. 15 "It was fascinating to see, when developing these ... taut skins, how the endeavor to build forms with a minimum of material gave rise to forms of great clarity and captivating beauty—forms no one would ever have dreamed of designing." 16

It was during preparations for the German pavilion at EXPO 67 that Otto and his colleague Larry Medlin developed a string of experiments in regard to geometric analysis of the potential forms of minimal surfaces. In the first experiments Otto opted for simple open frames. If the forms of minimal surfaces were to be geometrically analyzed using soap bubble models then the soap films would have to be more durable. The model would also need to be protected from drafts. A low-temperature, dust-free, and very humid environment would be required to prevent the soap films from rapidly drying out. The most important components of the minimal surface apparatus are: a chamber, air conditioning and humidity control, parallel light, a measurement grid, a projection umbrella, and a camera. Over the next circa fifteen years, the apparatus was further developed, in design terms and technically, and its functional range extended. f.6 At the latest here, it is clear that Otto's predilection for experiment was based not on a systematization of architecture in the narrow, natural scientific sense, but rather on the artistic interpretation of forms experimentally induced with the aid of architectonic parameters. The machinic experiment served not only the investigation of cause and effect but also, to an equal or even greater extent, to generate form as part of the design process. This dual objective was feasible only because Otto (the designer) consistently bowed (as an engineer) to the physical properties of tensile membrane surfaces, systematically researched their geometric properties, and used apparatus-based techniques to precisely fathom the potential of the construction method in hand. Here, techniques assume the role of "processing" and "making real" ideas. 17 Modeling techniques, drawing techniques, measurement techniques, and evaluation methods must accordingly be read as constituents of Frei Otto's innovative experimentation culture, which was rooted

f.5 Net model and overall model on the measuring table, German Pavilion for the EXPO 1967 in Montreal, 1965.

15 Cf. Frei Otto et al., "Forming Bubbles: A research project of the Institute for Lightweight Structures on minimal surfaces under the direction of Frei Otto," *Mitteilungen des Instituts für Leichte Flächentragwerke der Universität Stuttgart (IL)*, 18 (Stuttgart: Krämer, 1988).

16 Frei Otto and Peter Strohmeyer, "Zelte: Leichtbauweisen," *db deutsche bauzeitung*, 65 (1960), pp. 351–66; here p. 352.

17 Friedrich Dessauer, *Der Streit um die Technik* (Frankfurt/Main: Knecht 1956), pp. 225ff.

in the steady calibration of eye and hand, scientific observation, and mechanical dexterity—in short, in the self-assured fine-tuning of practical and intellectual capacities, whereby the act of architectural design propelled both individual knowledge production and collective debate of the discipline's future. For all their poetic fragility, the models in Frei Otto's work can certainly be described as exacting, for each reveals the tried and tested or as yet barely intuited insights of its maker. The most lasting impression on the mind's eye, however, is of an operative aesthetic that fuses the precision of scientific instruments with the allure of artistic intelligence.

f.6 Minimal surface apparatus with built-in wind tunnel, soap-film model in parallel light and camera.
→ 176/177

fischertechnik

Illustration Credits

Daniel A. Barber
Media Machines and Not-So-Utopian Futures, c. 1957
f.1 Eric Hodgins, "Power from the Sun," *Fortune*, 48, no. 9 (1953), pp. 130–5, 184–94; here p. 135.
f.2 M. King Hubbert, *Nuclear Energy and the Fossil Fuels* (Houston: Shell Development Co., 1956), fig. 11; Eugene Ayres, *Energy Sources: The Wealth of the World* (New York: McGraw-Hill, 1952), fig. 5, p. 164.
f.3 Schaible Company, *Kitchen of Tomorrow*, 1944. gta Archives, ETH Zurich (Bequest of Sigfried Giedion).
f.4 Photograph and explanatory drawing of the Olgyays' Thermohelidon. Jean Labatut Papers, Princeton University, Manuscripts Division, Princeton, New Jersey.
f.5 Victor Olgyay, *Design with Climate: Bioclimatic Approach to Architectural Regionalism* (Princeton, NJ: Princeton University, 1963), fig. 46.
f.6 George van Dyne, *Ecosystems, Systems Ecology, and Systems Ecologists* (Oak Ridge, TN: U.S. Atomic Energy Commission, 1966), fig 1.

Martin Bressani
Viollet-le-Duc's Organic Machine
f.1 [Eugène Emmanuel] Viollet-le-Duc, *Histoire d'un dessinateur: Comment on apprend à dessiner* (Paris: J. Hetzel, [1879]), fig. 60. Private collection.
f.2 Giovanni Alfonso Borelli, *De Motu Animalium* (Lugduni in Batavis: Apud Cornelium Boutesteyn, 1680), plate IV. Courtesy McGill University Library. Rare Books and Special Collections.
f.3/f.5/f.6 [Eugène Emmanuel] Viollet-le-Duc, *Dictionnaire raisonné du mobilier français de l'époque carlovingienne à la Renaissance*, vol. 5 (Paris: Morel, 1874), art. "Garde-bras," fig. 1; art. "Gantelet," fig. 8; art. "Grèves," fig.3. Private collection.
f.4 [Eugène Emmanuel] Viollet-le-Duc, *Dictionnaire raisonné du mobilier français de l'époque carlovingienne à la Renaissance*, vol. 6 (Paris: Morel, 1875), art. "Spallière," fig. 8. Private collection.

Carlotta Darò
Sound Conduits: Displaying the Architecture of Telecommunication
f.1/f.2/f.3/f.4 *Planning for Home Telephone Conveniences: For the Use of Architects, Engineers, Builders, Owners* (American Telegraph and Telephone Company, 1928), figs. 13, 18–20, 26, 27.
f.5/f.6/f.7 *Modern Telephone Service for the Home* (American Telegraph and Telephone Company, [192?]).
f.8/f.9 Advertising brochure for AT&T. Warshaw Collection of Business Americana, Archives Center, National Museum of American History, Smithsonian Institution.
f.10 Photograph from the George H. Clark Radioana Collection, Archives Center, National Museum of American History, Smithsonian Institution.
f.11/f.12/f.13/f.14 Advertising brochure for AT&T. N. W. Ayer Advertising Agency Records, Archives Center, National Museum of American History, Smithsonian Institution.

Kijan Espahangizi
Vitaglass: A Modern Boundary Technology between Laboratory Research, Architecture, Public Health and Environmentality in the 1920s and 1930s
f.1/f.6 Walter Eddy, "The Use of Ultra-Violet Light Transmitting Windows," *American Journal for the Public Health and the Nation's Health*, 18, no. 12 (1928), pp. 1470–9; here p. 1477, p. 1475.
f.2 Franz Mylius, "Die Prüfung der Glasoberfläche des Glases durch Farbreaktion," *Zeitschrift für Instrumentenkunde*, 9 (1889), pp. 50–7; here p. 52.
f.3 Karl Ford, "The Weathering of Glass Containers," *Journal of the American Ceramic Society*, 5 (1922), pp. 837–54, fig. 1–11b.
f.4 Eberhard Zschimmer, "Über neue Glasarten von gesteigerter Ultraviolett-Durchlässigkeit," *Zeitschrift für Instrumentenkunde*, 23, no. 12 (1903), pp. 360–3; here p. 363, fig. 2.
f.5 Das Jenaer Uviolglas 1908–1930, Schott Archive Jena 18/12.
f.7 Fritz Kahn, *Das Leben des Menschen: Eine volkstümliche Anatomie, Biologie, Physiologie und Entwicklungsgeschichte des Menschen*, vol. 4 (Stuttgart: Franckh, 1926), plate XXVII.

Sabine von Fischer
Tapping Machines: Listening to Difference, 1928–1956
f.1 Karl Willy Wagner: *Das lärmfreie Wohnhaus* (Berlin: VDI-Verlag, 1934), fig. 4.
f.2 a–c V. L. Chrisler and W. F. Snyder, "Transmission of Sound Through Wall and Floor Structures," *Bureau of Standards Journal of Research*, 2 (1928), pp. 541–59, fig. 4; Arnold Schoch, *Die physikalischen und technischen Grundlagen der Schalldämmung im Bauwesen* (Leipzig: Hirzel, 1937), p. 104; Leo L. Beranek, *Acoustic Measurements* (New York: Wiley, 1949), fig. 19.28.
f.3 Michael Bienert, *Moderne Baukunst in Haselhorst: Geschichte, Bewohner und Sanierung der Reichsforschungssiedlung in Berlin-Spandau* (Berlin: Berlin Story, 2015), p. 43.
f.4 a–b From Fritz Ingerslev, *Acoustics in Modern Building Practice* (London: The Architectural Press, 1952), p. 237; Werner Bürck, *Die Schallmessfibel für die Lärmbekämpfung*. 2nd ed. (Munich: Oldenbourg, 1960), p. 105.
f.5/f.6 Bernhard Wedler, *Baut ruhige Wohnungen* (Bad Godesberg: Bundesminister für Wohnungsbau, 1957), p. 3, cover.

John Harwood
Wires, Wirelessness and the Morality of Form in "The New Machine Art"
f.1/f.3 Alfred H. Barr Jr. and Philip Johnson, *Machine Art* (New York: Museum of Modern Art, 1934), cover, fig. 26.
f.2 Arthur Drexler and Greta Daniel, *Introduction to Twentieth Century Design from the Collection of the Museum of Modern Art* (Garden City, NY: Doubleday, 1958), fig. 132. Courtesy of the Museum of Modern Art, New York.
f.4 Charles Wheatstone, "On the Transmission of Musical Sounds through Solid Linear Conductors, and on their Subsequent Reciprocation," *Journal of the Royal Institution of Great Britain*, 2 (1831), pp. 223–38; here p. 231.
f.5 Robert J. Chapuis, *100 Years of Telephone Switching*, vol. 1: *Manual and Electromechanical Switching (1878–1960s)* (Amsterdam: IOS Press, 2003), chap. II-2, fig. 5.
f.6 *Encyclopédie ou Dictionnaire raisonné des sciences, des arts et des métiers: Recueil des planches.* 3rd ed. (Paris: Briasson; David; Le Breton, 1765), art. "Épinglier," plate II.

Julian Jachmann
Water / Space: Ledoux and the Disposition of Architecture and Machines
f.1/f.4/f.5/f.7 Claude-Nicolas Ledoux, *L'Architecture considérée sous le rapport de l'art, des mœurs et de la législation* (Paris: C. N. Ledoux and H. L. Perroneau, 1804), plates 3, 4, 6, 9.
f.2 Bernard Forest de Bélidor, *L'Architecture hydraulique, ou l'Art de conduire, d'élever et de ménager les eaux pour les différens besoins de la vie*, vol. 1, (Paris: Charles-Antoine Jombert, 1737), frontispiece.
f.3 Bernard Forest de Bélidor, *Architectura hydraulica* (Augsburg: Johann Georg Merz, 1743), vol. 3, chap. 4, plate 8. Werner Oechslin Library Einsiedeln.
f.6 Jean-Rodolphe Perronet, *Description des projets et de la construction des ponts de Neuilly, de Mantes, d'Orléans & autres*, vol. 2 (Paris: Imprimerie Royale, 1783), plate XXXIII, detail. Werner Oechslin Library Einsiedeln.
f.8/f.9 Charles-Etienne Briseux, *L'Art de bâtir des maisons de campagne* (Paris: Prault, 1743), vol. 1, plate 128; vol. 2, plate 142.

Susanne Jany
Making Buildings Work
f.1 Wilhelm Küsgen et al., *Handwörterbuch des Postwesens* (Berlin: Springer, 1927), p. 360.
f.2/f.3 Krinner, "Das neue Münchner Paketzustellamt," *Archiv für Post und Telegraphie*, 55, no. 7 (1927), pp. 173–81; here pp. 178, 179.
f.4 Felix Genzmer, *Wasch- und Desinfektionsanstalten*, Handbuch der Architektur, part 4, vol. 5, no. 4 (Stuttgart: Bergsträsser, 1900), p. 79.
f.5 Wilhelm Franz, *Fabrikbauten*, Handbuch der Architektur, part 4, vol. 2, no. 5 (Leipzig: Gebhardt, 1923), p. 215.

Reinhold Martin
Corporate Personhood: Notes Toward an Architectural Genealogy
f.1 Dartmouth Hall, Dartmouth College, 1784–91. Photograph, 1865. Courtesy of Dartmouth College Library.
f.2 Henry Dawkins, *A northwest prospect of Nassau Hall with a front view of the president's house, in New Jersey*, 1764. Courtesy of Princeton University Archives.
f.3 Herman Boye, *A Map of the State of Virginia*, 1827.
f.4 Edgar Allan Poe Room, *University of Virginia*. Photograph by Rufus W. Holsinger, 1915. Courtesy of Albert and Shirley Small Special Collections Library, University of Virginia.

Spyros Papapetros
"The Mechanization of Adornment" Revisited; or, The History of the Ornament-Machine
f.1/f.3/f.4/f.6/f.8 Sigfried Giedion, *Mechanization Takes Command: A Contribution to Anonymous History* (New York: Oxford University Press, 1948), pp. 349, 351, 358–9, 362–3, 386–7.
f.2 "Moultan Muslin, Manufactured by W. Govan and Sons," *Journal of Design and manufactures*, 1, no. 4 (1849), n.p.
f.5 *L'Architecture vivante,* 3, no. 10 (1925), plate 50.
f.7 Advertising brochure for *Mechanization Takes Command,* 1948. gta Archiv, ETH Zurich (Bequest of Sigfried Giedion).

Georg Vrachliotis
The Model as a Machine: Frei Otto's Architectonic Experimental Systems
f.1 a–c/f.2/f.3 a–b/f.4/f.5 Werkarchiv Frei Otto im Südwestdeutschen Archiv für Architektur und Ingenieurbau (saai), KIT Karlsruhe Institute of Technology.
f.6 "Forming Bubbles," *Mitteilungen des Instituts für Leichte Flächentragwerke der Universität Stuttgart (IL)*, 18 (Stuttgart: Krämer, 1988), pp. 329, 330.

Editors of this issue
Moritz Gleich
Laurent Stalder

Copy-editing
Jill Denton

Translations
Jill Denton (Stalder and Gleich, Weiss, Vrachliotis)
Thomas Skelton-Robinson (Erb, Schärer, Maurer)

Proofreading
Thomas Skelton-Robinson

Graphic concept and design
Christian Lindermann and Andreas Uebele
büro uebele visuelle kommunikation, Stuttgart

Production
Offsetdruckerei Karl Grammlich GmbH

Typeface
GT Eesti

Paper
MaxiScript 90 g/sm
MaxiScript 190 g/sm

Bibliographic information published by the Deutsche Nationalbibliothek
The Deutsche Nationalbibliothek lists this publication in the Deutsche Nationalbibliografie; detailed bibliographic data are available on the Internet at http://dnb.dnb.de.

ISBN 978-3-85676-363-3

ISSN 2504-2068

gta Verlag

ETH zürich